✓ 2002

M. David Stone
Award-winning writer and
Contributing Editor, PC Magazine

Ron Gladis
Award-winning
video producer
and photographer

D1033457

Faster Smarter

Digital Photography

Take charge of your digital camera and images—
faster, smarter, *better*!

PUBLISHED BY
Microsoft Press
A Division of Microsoft Corporation
One Microsoft Way
Redmond, Washington 98052-6399

Copyright © 2003 by M. David Stone and Ron Gladis

All rights reserved. No part of the contents of this book may be reproduced or transmitted in any form or by any means without the written permission of the publisher.

Library of Congress Cataloging-in-Publication Data
Stone, M. David.
    Faster Smarter Digital Photography / M. David Stone, Ron Gladis.
      p. cm.
    Includes index.
    ISBN 0-7356-1872-0
    1. Photography--Digital techniques--Handbooks, manuals, etc. 2. Digital
cameras--Handbooks, manuals, etc. 3. Image processing--Digital
techniques--Handbooks, manuals, etc. I. Gladis, Ron. II. Title.

TR267 .S76   2002
778.3--dc21   2002033738

Printed and bound in the United States of America.

1 2 3 4 5 6 7 8 9   QWE   8 7 6 5 4 3

Distributed in Canada by H.B. Fenn and Company Ltd.

A CIP catalogue record for this book is available from the British Library.

Microsoft Press books are available through booksellers and distributors worldwide. For further information about international editions, contact your local Microsoft Corporation office or contact Microsoft Press International directly at fax (425) 936-7329. Visit our Web site at www.microsoft.com/mspress. Send comments to *mspinput@microsoft.com*.

Microsoft, Microsoft Press, Picture It!, Windows, and Windows NT are either registered trademarks or trademarks of Microsoft Corporation in the United States and/or other countries. Other product and company names mentioned herein may be the trademarks of their respective owners.

The example companies, organizations, products, domain names, e-mail addresses, logos, people, places, and events depicted herein are fictitious. No association with any real company, organization, product, domain name, e-mail address, logo, person, place, or event is intended or should be inferred.

Acquisitions Editor: Hilary Long
Project Editor: Aileen Wrothwell
Series Editor: Kristen Weatherby

Body Part No. X08-99936

To Marie and Mariah

Table of Contents

Part I: The Basics: What You Need to Know

The part title says it all: the chapters in this part will tell you everything you need to know, whether it's what you need to know to pick the right camera, understand the features in the camera you picked, or take better pictures. They'll also cover such practical issues as how to take best advantage of your camera's memory and how to make your batteries last longer. Armed with the knowledge in these chapters, you'll be all set to take great pictures.

Part II: Getting Creative and Cutting Loose

One of the advantages of taking pictures on a digital camera instead of using film is that they are so easy to edit, both to make them more interesting and to fix problems. The chapters in this part show how to crop and resize your photos, remove flaws, add artistic effects, adjust colors, use your photos as screen saver images, and more—including how to stitch photos together to create a panoramic image, and, more important, how to take the photos so you can stitch them together successfully.

Part III: Sharing Your Photos

With digital photography, you have a wide range of choices for how to share your photos. This section covers most of the possibilities, including printing your own photos, having the photos professionally printed, posting them on a Web site, and e-mailing them. It also covers such issues as inserting pictures into documents and creating a slide show to view your photos on your computer monitor or TV. It also discusses the best choices for transferring slides to videotape that you can play in your VCR, and to CD discs that you can play in your DVD player.

Acknowledgments

One of the benefits of writing books like this one is that you get to play with toys without having to buy them (even if you do have to return them when you finish, alas). We'd like to thank Nikon, Casio, Olympus, Logitech, and Epson for providing the cameras we used for specific examples in this book. Thanks, too, to Nikon for providing the picture of the Nikon Macro Cool-Light SL-1 that you'll find in Chapter 3.

We'd particularly like to thank Epson for providing two Epson PhotoPC 3100Z cameras and the Epson Stylus Photo 785EPX printer. We used the cameras to take most of the photos in this book. We used the printer both as an example in the book, and as our primary printer for testing while writing. Special thanks go to Kylie Ware for arranging the equipment loan from Epson, along with plenty of photo paper for printing, and to Karen Thomas, who not only arranged the loan from Olympus, but also provided a list of photography-related Web sites that saved us hours of research time.

We'd also like to thank SanDisk for providing high capacity memory cards in several formats, as well as a card reader that not only served as an example in a photo, but also came in handy for moving photos to our computers. And we'd like to thank the companies that provided the latest versions of software for our examples, notably Adobe, Jasc, Scansoft, ECI, and Microsoft.

Closer to home, we'd each like to thank our families, who tolerated the odd hours we both had to put in to get this book done on time. And thanks to our agent, Claudette Moore, who did her usual, excellent job of making sure the project would go from an idea under discussion to a finished book.

Finally, we'd like to thank everyone at Microsoft Press who made this book happen and who helped make it better, starting with Jeff Koch, who began the discussions about the book, and Hilary Long, who took over. Our editor, Aileen Wrothwell, worked hard to make sure all the pieces would come together. And we'd also like to thank those we've never dealt with directly, or know only by way of comments on our manuscript. Most notable among these is our technical editor, Tom Keegan (who is clearly brilliant, given that his biases match ours and his suggestions were invariably on target). Just as important, are those whose copy editing, graphic design, layout, and other work added to the quality of the finished book.

Introduction

Writers have long lamented that if you introduce yourself at a social occasion as a writer, the most common response is, *I've got a story that would make a great book*. After which, you have to listen to the story. Photographers are rapidly catching up with a most common response of their own.

What do you do?
I'm a photographer.
Really! Tell me...should I buy a digital camera? Are they any good yet?

The answer, of course, is a definite maybe.

This Book Could Be for You

If you're still unsure whether to take the plunge into the world of digital photography and need to be convinced that a digital camera really is worth getting (or maybe that you should wait awhile longer), this book will show you what a digital camera can do. If you're ready to buy but aren't sure what to get, this book will help you pick the right camera. If you've recently bought a digital camera and are just learning to use it, this book will help speed you through the learning curve. But most of all, this book will show you what you can do with your digital photos once you've taken them, and guide you into some nooks and crannies that you might not find on your own.

Of course digital photography is not just about digital cameras.

For a start, it's also about photography. No matter how much or how little you know about digital cameras as hardware, your photographic expertise may lie anywhere on the scale from point-and-shoot basics to a deep knowledge of shutter speeds, f-stops, and peculiarities of one type of film compared to another. We make no assumptions about how much you know, except that it's less than what a professional photographer is likely to know. It may be a lot less—limited to knowing how to put the film in the camera, point, shoot, and take the film out when you're done. If so, this book will teach you photography basics as you learn about digital photography. If you already know the basics, you'll see how what you already know applies in the context of digital cameras.

You'll also learn some things that are specific to digital cameras. For example, you need to know how to maximize battery life so you don't have to carry a ton of batteries with you every time you go out with your camera. And you need to know your options for offloading photos from the camera storage to make room for more pictures. (Buying extra storage is the obvious choice, but not the only one.)

Digital photography doesn't stop once you've taken the pictures either. In fact, the majority of this book will cover things you can do only after you've taken your photos.

You can edit the pictures in various ways. This book will show you how to:

- Crop out unwanted elements
- Size your photographs to meet your needs
- Digitally enhance—or flat out change—reality
- Reformat
- Add text
- Turn your photo into a greeting card

You can share your photos with others. This book will discuss how to:

- E-mail photos
- Share photos on a Web site
- Make sure files are small enough to make e-mailing or viewing on the Web practical
- Add photos to your letters, newsletters, genealogy charts, and other documents

You can view your photos without needing a printed photo. This book will show you how to:

- Best view your photos on a monitor
- View your photos on a TV screen to create a digital slide show
- Store and view your photos on a personal digital assistant (PDA)—wallet photos can be a thing of the past

You can print your photos on a computer printer. This book will cover how to:

- Choose a printer for photos
- Print your photos in wallet size, life size blow-up, or anything in between

- Choose the right paper to give you the best photos possible, or give you the right compromise between photo quality and cost

You can also store the photos in what amounts to digital albums. This book will cover such issues as:

- Storage options—like Zip disks, CD-R discs, and hard disks
- Programs and Web sites that let you store and manage photos
- How to choose the right resolution and compression so your photos will take up the least possible disk space without losing picture quality

To help you get the most out of the book, you'll find helpful elements within the chapters:

- Lingo notes to indicate a technical term or some jargon you should be aware of
- Tips that provide helpful hints or tricks
- Notes that provide just a bit more information
- Cautions that warn you about potential pitfalls
- Try This! exercises that provide a practical application of the topic just covered in the text
- Sidebars that have technical or background information that you don't really have to know but might find interesting

System Requirements

We make no assumptions about your computer system except that it's appropriate for the camera you've bought and meets whatever requirements that the camera manufacturer recommends. That said, however, we will make recommendations throughout the book based on features you *may* have or can get. In talking about ways to move photos from your camera to your computer, for example, we will mention some possibilities that require a PC Card slot in the system. If you have a notebook, you probably have a PC Card slot. If you have a desktop system, you probably don't, but you can add one (and we'll explain how).

Support

Every effort has been made to ensure the accuracy of this book. Microsoft Press provides corrections for books at the following address:

http://mspress.microsoft.com/support/

If you have comments, questions, or ideas regarding this book, please send them to Microsoft Press via e-mail to:

mspinput@microsoft.com

or via postal mail to:

Microsoft Press
Attn: Faster Smarter Series Editor
One Microsoft Way
Redmond, WA 98052-6399

You can also contact the authors directly at digicambookcomments@comcast.net with any comments or suggestions.

Please note that product support is not offered through the above addresses.

The Basics:
What You Need
to Know

Digital photography is different from film photography in some important ways. The first part of this book will tell you what you need to know to get started, no matter how much you already know about photography on film. The chapters in this section compare digital photography and film photography to help you understand the strengths and weaknesses of each (and help you decide whether it's time to go digital if you haven't already made that decision). They discuss features and types of cameras to help you pick the right camera or help you confirm that you did. They also discuss what you need to know to take pictures—both the settings you need to worry about in the camera and the various issues you should consider to make sure you take pictures you'll want to keep.

Most important, this part discusses considerations that apply only to digital cameras, like how to extend your battery life and choices for how to move the pictures to your computer. By the time you work your way through this part, you'll be able to claim that you know your way around digital photography. You'll also be ready to take advantage of all the suggestions for creative use of your photos that we'll get to in the rest of the book.

Chapter 1

Everything's Coming Up Digital

The proverbial handwriting isn't just on the wall, it's emblazoned in neon lights that you'd have to be blind to miss: digital cameras will unquestionably take over from cameras that depend on film, just as the digital technology of CDs took over from vinyl records, and just as digital cell phones are taking over from analog cell phones. It may not happen quickly, but it will happen.

Most professional photographers today are already using digital cameras at least sometimes. So are increasing numbers of serious amateurs (often called *prosumers*) and people who just want to take snapshots.

Lingo *Prosumer* is shorthand for consumers who know enough about photography so they could pass themselves off as professional photographers. It's essentially synonymous with serious amateurs.

It's important to understand that digital cameras aren't taking over the world of photography because digital images are better than film images, or even that they are just as good. You can make a strong case that they're not. However, they have reached the point of being good enough for most purposes. That's an important distinction, and one that's worth looking at more closely.

Digital Cameras: The Professional Photographer's Choice (at Least Sometimes)

If we had to pick two major trends to bet on as most likely to take over in any technology that they apply to at all, the choice would be easy. The first half of our bet would be on the trend toward color. The second half would be on the trend toward digital technologies. As it happens, both are critical elements for digital photography, and both trends have already taken over as much as they need to for digital photography to flourish. Otherwise we wouldn't be writing this book, you wouldn't be reading it, and professional photographers wouldn't be using digital cameras.

Color first: the human brain seems hard-wired to prefer color. Movies, still photographs, televisions, computer screens, and more all started out in black and white—or, more precisely, *monochrome*—formats, and all have moved (mostly) to color. With any of these technologies today, black and white is used only to get a specific artistic or conceptual effect—like showing the first scenes in *The Wizard of Oz* in monochrome to make the point that (on the surface at least) dull, old monochrome Kansas doesn't hold a candle to glorious, full-color Oz.

Lingo   *Monochrome* means one color, but not necessarily black and white. Lots of early computer monitors showed white text on a black background, but others showed green text or amber text. All were monochrome.

Of course, part of the reason these technologies started out with black and white was that the black-and-white versions were easier to produce, not to mention cheaper. The color versions of each one started getting popular as the cost for color went down, but the driving force behind the switch remains the preference people have for color over black and white. If digital cameras were available for taking black-and-white pictures only, people would be staying away from digital photography in droves (and odds are that you wouldn't be reading this book).

Try This!   You probably don't need us to tell you that color is generally more visually interesting than black and white, but in case you have any doubts, there's an easy way to prove it. If you already have a digital camera, it probably came with a photo editing program. Or, you may have a program that came bundled with a scanner or with some other software. If the software is installed, and you're already comfortable using it, look for a feature in the program that will convert an image from color to black and white. (Don't get frustrated if you can't find it; not all programs offer the feature. And if you aren't comfortable using the software yet, skip this for now. You'll learn how to do this sort of thing in the second half of this book.) Then load some color photos in the program; if you haven't taken any photos yet, look for some sample images that came with the program.

Convert each photo to black and white. (Be careful not to save any of the modified images, which would overwrite the color versions on disk.) Your results will vary depending on the images you pick, and you may find that some of the photos actually gain visual interest in black and white (a point we'll discuss in Chapter 4 in the section "Black and White Versus Color"). But you should find that the vast majority of photos look much more appealing in color.

You've Got to See Your Pictures After You Take Them

From a photographer's point of view—and that includes everyone from professionals to the most occasional amateur—the two most important technologies that have switched to color are computer monitors and computer printers. And this is a case where the benefit of having color available in both technologies is much greater than the sum of the parts.

If you couldn't get good quality color output from either monitors or printers, it wouldn't matter if you could take great pictures with a digital camera; you wouldn't have any way to see them. If you could get good-quality color from monitors only—which was true for a long time—you'd be able to see your pictures on screen, edit them, and do things like e-mail them or show them on the Web, but you wouldn't have any way to print them. If you could get good quality color from printers only, you'd be able to print your pictures, but you wouldn't be able to edit them.

Having good color quality on both monitors and printers opens up more possibilities. Not only can you do everything that you could do with a color monitor alone plus what you could do with a color printer alone, but you can do more. The combination gives you what amounts to your own sophisticated photo lab tucked inside your computer. Take a picture, and you can print it, e-mail it, or post it on a Web site as is. Or you can enhance it, add graphic elements or text, or crop it (cut off part of the image) and resize it, as in Figure 1-1, and *then* print it, e-mail it, or post it.

Note Practical considerations, alas, mean that the photo samples in this book will all be monochrome. Little is lost, however, because there is little we would be able to show in color that we can't show effectively in monochrome format as well.

Figure 1-1   The top right picture and the two bottom pictures are all edited versions of the top left original.

For a professional photographer dealing in, say, portraits or corporate bro-chures or annual reports, this ability to quickly and easily modify an image on his or her computer screen instead of working with messy chemicals, sending the job out to a photo lab, or taking the time to scan the photo into digital form, not only saves time and money, but gives better control over the final product.

For a nonprofessional, it opens the door to doing things that you would never have done with film—no, make that never would have even *thought* of doing with film.

A Bit about the Digital World

The driving force pushing us all to digital technologies comes from a very dif-ferent direction than the drive to color. In truth, people's perceptions aren't wired for digital anything the way they are wired for color. To the contrary, they are wired for *analog* technologies, which allow continuous change rather than the discrete steps that digital technologies offer.

Lingo   *Analog* technologies change whatever factor they change—color, brightness, audible tone—in a continuous way, rather than in clearly defined steps.

Consider a digital clock versus an analog clock as shown in Figure 1-2, for example. The digital clock shows hours and minutes. It can show you when the time is, say, 12:13 or 12:14, but it can't show anything in between. Analog clocks—by which we mean real analog clocks, preferably with a sweep second hand, not the re-creations drawn on the face of some digital clocks—can't skip the in-between times. To get from 12:13 to 12:14, the second hand has to trace out all the seconds in between.

Figure 1-2   Digital clocks show time in discrete steps, jumping from one second to the next. Analog clocks show time as falling along a continuous circle.

That's the essential difference between digital and analog technologies, and it's really all you need to know about the differences: Digital technology changes things in steps. Analog technology changes things continuously.

This difference is so obvious and so simple that it's easy to miss how important it is and in how many different ways it comes into play. Another example or two will help. (We've made a point of picking examples that relate directly to digital photography and will give you a basic understanding of some of its possibilities and weaknesses. They will also help you understand what we mean when we say that digital photography has gotten good enough.)

We'll start with a simple case. Suppose that you want to draw a diagonal line on a piece of paper, going down, from left to right. The technology you're using—pencil and paper—is analog in nature. (You probably don't usually think of pencil and paper as technology, but it is. Consider how much trouble it would be to run a civilization without it.) That means you can drag the pencil point along the paper—preferably using a ruler to get a straight line—and wind up with a nice, smooth diagonal line, like the one in Figure 1-3.

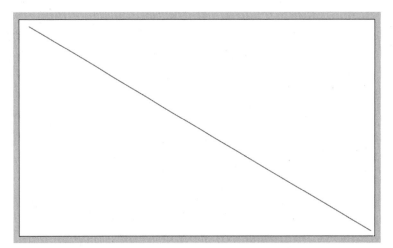

Figure 1-3   An analog technology like a pencil will give you a smooth line, like this one.

If you try the same thing with a digital technology, drawing a line with a drawing program on a computer monitor for example, the line comes out in steps—one step right, one step down, one step right, and so forth, like the one in Figure 1-4. The result is anything but smooth.

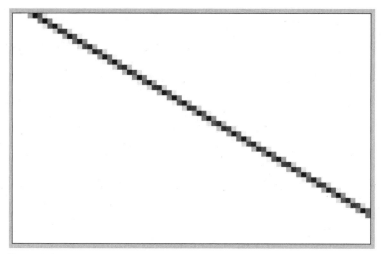

Figure 1-4   A digital technology will give you a staircase step effect for a diagonal line.

The steps that use lighter grays in this line are meant to help blur the steps and give the line a smoother looking edge than if it used black only. Even with those blocks of light gray, however, this second example of a diagonal line is not an acceptable substitute for a straight line—at least, not to the human visual system. But now comes a subtle irony.

In the real world, at the most basic physical level, everything comes in steps. The graphite in the line laid down by the pencil and, for that matter, the paper itself, both have a built-in graininess. More than that, they are grainy on all sorts of levels—from a roughness in the paper surface visible even with a good magnifying glass, to individual molecules and atoms, and even to levels below that. (And that's as far as we will venture into physics).

The point is that if the steps are small enough, you can't see them. Or, to put it another way: if a digital technology breaks the steps down far enough, it can effectively erase the difference between digital and analog technologies. As proof of that point, note that Figures 1-3 and 1-4 both show the same line drawn in the same program. The difference is that we zoomed in on the line to get the second figure and make the steps far larger and more visible.

Digital Cameras and Resolution

Probably the most obvious area where the idea of digital steps comes into play for digital cameras is resolution—the number of pixels the camera uses. If you know anything about digital cameras at all, it's probably that they are rated in megapixels and that more megapixels is better. A 3-megapixel camera offers better pictures than a 2-megapixel camera, and a 6-megapixel camera offers better pictures still. More megapixels also makes for a more expensive camera. In case you're not too clear on what a megapixel is, however, we'll start with some basics.

Digital photos are similar to mosaics, which create pictures from the arrangement of small tiles. In a mosaic, the tile is the smallest unit of the picture. In a digital photo, you use *picture elements*, usually abbreviated as *pixels*, and the pixel is the smallest unit of the picture. Whether you're dealing with tiles or pixels, each one has a specific color, and it's the arrangement of the tiles or pixels that creates the picture.

Lingo   The smallest element of a digital picture is a *picture element,* which is usually abbreviated to *pixel.*

A *megapixel* is a million pixels, more or less. (Actual cameras tend to round their claims to the nearest megapixel, or sometimes the nearest tenth of a megapixel.) The exact numbers aren't important. What you need to know for the moment is that the pixels are arranged in a rectangular area—so many pixels across by so many pixels down. Multiply the number across by the number down, and you'll wind up with the maximum number of pixels that the camera delivers. A camera that shoots photos at 1600 × 1200 resolution, for example, delivers 1,920,000 pixels (and would usually be called a 2-megapixel camera).

The important point is that any given camera, and, therefore, any given photo the camera takes, is limited to some maximum number of pixels. That number stays the same whether you print the picture at 3 × 4 inches, 8 × 10 inches, or big enough to fill a billboard. It also stays the same whether you show the photo on a monitor with 640 × 480 resolution, or expand it to fill a monitor with 1600 × 1200 resolution.

The potential problem—which is where the issue of steps comes in—is that no matter what size you print the photo or expand it to on screen, the size of the pixels stays the same relative to the full photo size. (This is a bit of an oversimplification, because there are ways around the issue, but it's essentially right.)

Print the picture at a large enough size, or expand it to a large enough size on screen, and the photo may get *pixelized*, so you can see the individual pixels.

Lingo   A *pixelized* photo is one that's enlarged enough so you can see the individual pixels.

Figures 1-5 through 1-7 show the same photo as it looks on screen at different levels of zoom. (We captured the screens using a software utility and saved them as files.) In Figure 1-5 the photo is set up at 100 percent zoom, which means each pixel on the screen uses one pixel in the photo. There's no pixelization at all.

Figure 1-5   A photo as it appears on screen at 100 percent, using one pixel on screen for each pixel in the photo.

In Figure 1-6, we zoomed in to 200 percent, which is equivalent to taking a photo that looks fine at 3 × 5 inches and enlarging it to 6 × 10 inches—or taking a photo with 640 × 480 resolution and enlarging it to fill a monitor screen running at 1280 × 1024 resolution. By doubling both the height and width of the image, you can begin to see a hint of the pixels, particularly in the whiskers in the lower left of the picture.

Figure 1-6   The same photo as it appears at 200 percent, showing some hints of pixelization, particularly in the whiskers at the bottom left .

In Figure 1-7, finally, we zoomed in to 400 percent, doubling the size of the picture again. This is equivalent to taking a picture designed for printing at 3 × 5 inches, deciding to focus on an area about half that height and width, and then printing it at 8 × 10 inches. Here you can clearly see the small rectangular pixels in the eye and in the cat's surrounding fur.

Figure 1-7   At 400 percent the pixelization in this photo is much more obvious.

Once again, the basic point here is that as long as the steps are small enough, as in Figure 1-5, you won't see them. As the steps get larger, they get more visible. If we had gone to still greater enlargement than in Figure 1-7, the pixels would be even more obvious.

One of the most significant improvements in digital cameras over the last few years has been the welcome growth in pixel count at increasingly affordable prices. It wasn't that long ago that to be affordable, a camera had to offer half a megapixel or less—barely suitable for taking pictures to see on screen or print at 3 × 5 inches. Today's inexpensive cameras offer enough pixels to be suitable for printing at 8 × 10 inches or even larger. More important, most offer enough pixels to let you crop an image to eliminate parts you don't want, and then enlarge the remaining part of the picture to a reasonable size without pixelization. That's one of the hurdles digital cameras had to overcome to make them good enough to be worth using.

About Color

The idea that if you use small enough steps you can effectively provide continuous, analog-like results applies to color as well. As we've already pointed out, color is one of the most important features in digital photography. Without it, digital cameras could take photos with as many pixels as you'd ever need, and they'd still have limited appeal. Color is also one of the features that gets *digitized*, or converted into digital information. And if you want to take best advantage of color in your photos, it helps to know a little about how digital color works. Before we can talk about digital color, however, we have to talk a little about color in general.

Lingo   *Digitized* information is information that's been put into digital format. The process is called *digitizing*.

Color is a far more complicated subject than most people realize. Some people spend their entire careers as color scientists working for companies that produce things like film, scanners, and digital cameras. But we're not going to get deep into color theory here. We're simply going to touch on some highlights you'll find useful to know.

Mixing Colors: How Cameras and Monitors Work

One thing most of us learned early on in life, when finger painting was the height of our artistic achievement, is that when you mix two colors together, you get a third color. Somewhere along the way you may also have paid enough attention in an art or physics class to learn that if you start with the right colors, you can use just three colors to create all the other colors you can see.

There happen to be different combinations of colors that will work in different situations. For printers, the preferred starting colors, or *primary colors*, are cyan, yellow, and magenta. (Cyan is a light blue; magenta is a purplish red.) For digital cameras—as well as monitors, scanners, and film cameras, for that matter—the primary colors are red, green, and blue.

Lingo   You can create any color from three *primary colors*, which, for a digital camera, are red, green, and blue.

You may or may not be aware of how color televisions and computer monitors trick the eye into seeing color. (Both work the same way.) There are some variations on the theme, but, basically, the screens have sets of red, green, and blue phosphor dots, arranged as in Figure 1-8.

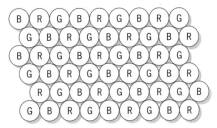

Figure 1-8   Monitors create all their colors with just red, green, and blue phosphor dots.

Monitors and televisions can't mix colors the way you used to mix finger-paints, because the dots are permanently in place. What monitors and TVs do instead is make the different color dots glow at different intensities to add more of one primary color and less of another. The dots are close enough together so that instead of seeing separate dots, the eye mixes the colors together to come up with the colors you see.

Try This!   To get a better feel for how monitors and televisions mix colors, take a look at the phosphor dots themselves. The dots on most television screens are big enough so you can see them with the naked eye if you get close enough to the screen. For most computer monitor screens, you'll need a magnifying glass.

If you have a magnifying glass handy, make sure your computer is showing an image that includes some color areas and some white areas, then take a look at each area. Depending on the particular monitor, the phosphors may be arranged in dots as in Figure 1-8 or they may have a similar arrangement with different shaped dots, or the phosphors may be arranged in stripes that run from the top of the screen to the bottom in repeating patterns of red, green, and blue stripes. Whatever the arrangement of the phosphors, you should be able to see the individual dots, in just three colors, when you look through the magnifying glass.

Make sure that you take a look at a white area in particular. When dealing with light, the three primary colors red, green, and blue mix to form white. Actually seeing that the white areas are made up of red, green, and blue is the best way we know to convince yourself that this really works.

If you don't have a magnifying glass handy, you can try this with a television screen instead. However, it's harder to look at an area with a particular color on a television, because the picture keeps moving, and any given area is constantly changing color.

Now, pretend for a moment that color digital cameras don't exist and your boss has given you until next Tuesday to invent one. One thing you could do is link three monochrome cameras together so they can take the same picture at the same time, then put a red filter in front of the first camera, a blue filter in front of the second, and a green filter in front of the third.

When you take the picture, you'll have three different monochrome pictures, all in black and white. However, they'll look different because the first camera saw only the red in the scene as in Figure 1-9, the second camera saw only the green as in Figure 1-10, and the third camera saw only the blue as in Figure 1-11. (We didn't use three cameras for these figures. We'll explain how we got them a little later.)

Figure 1-9   In this red component of a color photo the barn is lighter than it is in the green and blue components because the barn is red.

Figure 1-10   In this green component the grass is lighter than it is in the red and blue components because the grass is green.

Figure 1-11   The blue component of the same photograph has nothing particularly notable in it.

Once you have these three images, all you need is some software that will show all three together at the same place on your screen, one effectively on top of the other, with each one lighting up just one set of phosphor colors—red, green, or blue. The first one needs to show as a monochrome red image (red phosphor dots only), the second as a monochrome green image (green phosphor dots only), and the third as a monochrome blue image (blue phosphor dots only). Since your eye will mix the colors together, you'll wind up with a full-color image.

And that's basically how digital cameras (and monitors) work. There are, of course, a few refinements you can add—like taking the red, green, and blue pictures all at once in one camera—but we'll leave those details to the engineers.

Color as Shades of Gray

Not so incidentally, the way we got the red, green, and blue components of the photo in Figures 1-9 through 1-11 was almost the reverse of the process we described to create a color image. We started with a full-color photo.Then, for each of the figures, we edited the original image in a photo editing program to eliminate the other two colors and show the monochrome image in black and white.

If the process sounds mysterious right now, don't worry about it. This is precisely the sort of fun and games with photos that we cover in the second part of the book. There are really only two pieces of information that matter:

- First, by combining the three images together, you get the full-color image.

- Second, the black-and-white versions of each image contain all the color information you need for the one color they represent: red, green, or blue. Just add the right color for each, and combine the three images.

This ability to start with three black-and-white versions of a color scene and end up with a full-color image has led people who work with this stuff every day to talk about color in terms of *shades of gray*. So if someone says that a camera or monitor can produce, say, 256 shades of gray, what they mean is that it uses 256 shades of red, 256 shades of green, and 256 shades of blue. And, by the way, if you're interested in a black-and-white image, it can also give you 256 shades of actual gray.

Lingo   *Shades of gray* is shorthand for talking about the number of shades of red, shades of green, and shades of blue a monitor or camera uses.

This idea of referring to shades of colors as shades of gray is a useful shorthand, and it's one we'll use, too. However, there's more to it than just a shorthand description; there's a recognition that there is a certain equivalence between black-and-white images and color images. More precisely, if there are too few shades of gray to reproduce a color image well, the same flaws will show up in a black-and-white image with the same number of shades.

That's an important point: much of what's true for black and white is also true for color. And that means you can simplify discussions about color by talking about black-and-white images and shades of gray instead of having to talk about shades of colors and trying to visualize how the three primary colors interact with each other. More important for our purposes—since we're limited to using monochrome images in this book—it means the examples you'll see in this book in monochrome apply just as well to color images.

Digital Color

We've snuck it in here without any fanfare, but you may have noticed that suddenly we're back to talking about steps. When we talk about 256 shades of gray, what we're talking about is 256 individual steps: black, white, and 254 additional, discrete steps in between.

The human eye doesn't see the real world that way, of course. When you look at, say, a cloudless sky, the shade of blue changes between the horizon and straight overhead, but it doesn't change in sudden, discontinuous steps; it changes in a gradual, continuous way, much like the grays in the monochrome image in Figure 1-12 shade continuously from one to the next.

Figure 1-12   This graphic shades from black to white, changing in a continuous way.

We're using a graphic image rather than a photo in Figure 1-12, because it lets us control the range of shades. In this case, the range, or *gradient*, goes all the way from black on the left to white on the right. However, the same sort of gradient shows up in all sorts of real-world objects and the photos of those objects. We already mentioned the blues in the sky. Flesh tones are another key area where shades change gradually and continuously. Figure 1-13 shows examples of both.

Lingo   A *gradient* is an area that changes shades—or, at least should change shades—in a continuous way, from one color or shade of gray to another.

Figure 1-13   In the real world, colors (reduced here to shades of gray) often change in a continuous way.

Obviously, digital photos don't (in general) show steps in this sort of image. If they did, you wouldn't expect many people to buy digital cameras. As a point of reference, and to give you a sense of what the steps look like with various numbers of steps, Figures 1-14 through 1-20 show a graphic gradient like the one in Figure 1-12 as it would show with 2 to 128 steps.

Figure 1-14   A gradient from black to white using only 2 steps: black and white.

Figure 1-15   The same gradient, using 4 steps.

Figure 1-16   The same gradient again, using 8 steps.

Figure 1-17   The same gradient with 16 steps.

Figure 1-18   With 32 steps, the steps start getting more subtle.

Figure 1-19   With 64 steps, the steps may not show in the final printed output.

Figure 1-20   With 128 steps, the steps often will not show.

It's hard to predict exactly how each of these samples will look in the final printed version of this book compared to how they look on screen as we are writing, but we expect that the steps will be obvious at least up to Figure 1-18—the version that shows 32 steps. Figure 1-19, with 64 steps, may or may not show the steps when printed. As viewed on screen, however, the steps are subtle, but easily visible. Figure 1-20 probably won't show any steps when printed.

That, of course, is the point of this exercise: if you make the steps small enough, you won't be able to see them.

The obvious question is, how small do the steps have to be? The long answer is in the next section, "Shades, Colors, and Bits." The short answer is 256 shades of gray, even though you can sometimes get by with less. The gradient we started out with in Figure 1-12 uses 256 shades of gray, which means it should certainly not show any steps.

By the way, don't confuse the number of shades of gray with the number of colors. If you've ever used 256-color mode on your computer, you may have noticed that it's not suitable for viewing photos. The shading is nowhere near continuous, and you can easily see the steps as discrete colors with sharp boundaries going from one shade to another. That's because 256 colors is not the same thing as 256 shades of gray. It's actually close to 6 shades of gray. (The next section also explains the relationship between these two approaches to counting the number of colors.) When a photo shows obvious steps in the shading, it's called *posterized*.

Lingo   A *posterized* image shows sudden changes in shading that should change gradually. This *posterization* can be added as an artistic effect, or can show up accidentally as a flaw in an image.

To give you a sense of how some of the choices for the number of shades of gray translate into photos, Figures 1-21 to 1-23 show the two sample photos with four different levels for shades of gray. You'll notice that with 8 shades of gray, in Figure 1-21, the steps are extremely obvious. You might do something like this for artistic effect. In fact, most photo editors include a posterize effect that does exactly this. However, you wouldn't want all your photos to look this way.

Figure 1-21   Our sample photos with only 8 shades of gray.

Much the same comment applies to Figure 1-22, with 32 shades of gray. However, the steps here are far less obvious and may not even be visible in the final printed form of this book, although they are easily visible on screen with the photos sized to show each pixel in the image as one pixel on screen.

Figure 1-22   The same photos with 32 shades of gray.

In the top two images in Figure 1-23, at 128 shades of gray, the individual steps aren't visible on screen, and probably won't be visible in the printed version of this book either. If so, you won't see any difference between the top two photos and the bottom two, which use 256 shades. However, it's always safer to stay with the higher number of shades of gray.

Figure 1-23   The same photos again with 128 shades of gray in the two photos on top, and 256 shades in the two photos on the bottom.

Shades, Colors, and Bits

As we mentioned in the last section, the number of shades of gray is not the same thing as the number of colors. However both numbers will tell you the *color depth*, or how many colors are available in a given digital scheme, with more colors translating to a greater, or higher, color depth.

Lingo   The *color depth* tells you the number of colors available, with more colors giving you greater, or higher, color depth.

There's a straightforward relationship between these two ways of giving color depth, and it's useful to know how to translate from one way of counting colors to the other. A slight complication, however, is that there's a third way of counting colors too, and we need to cover that third way first.

If you've ever adjusted the resolution or other video settings on your computer, you may have changed the setting for the number of colors as well, or at least noticed that you can change the number if you want to. Some games, for example, require that you switch to 256-color mode, after which you may want to switch back to something else.

The choices you have available—and what they are called—depend on your video card, your monitor, your version of Microsoft Windows, and the resolution you're using. They also depend on where you look, since some choices are hidden away where you won't usually come across them.

Keeping that hedge in mind, a typical list includes 16 colors and 256 colors at the low end of the choices, and then switches to talking about the number of bits instead of the number of colors, most often with choices for 16 bits and either 24 bits or 32 bits or both. Adding still more room for confusion, the 16-bit choice is sometimes listed as high color or medium, and both the 24-bit and 32-bit choices are sometimes listed as true color or highest color.

The good news is that all these different approaches to counting colors—number of shades of gray, number of colors, and number of bits—are reasonably easy to sort out.

Bits and Colors

First, it helps to understand what a bit is. In addition to being digital, computers are binary, which means they represent everything with just two numbers: 0 and 1. Each of these numbers is a binary digit, which is usually abbreviated to *bit*.

Lingo   *Bit* is short for *binary digit*, which is a 1 or 0 in the binary counting system that computers use.

Computers keep track of colors (or, more precisely, programmers tell computers how to keep track of colors) by assigning some arbitrary number of bits to create a code for each color. This works something like Morse code, which uses dots and dashes, and assigns an arbitrary pattern of dots and dashes to each letter of the alphabet. For example, in Morse code, the pattern • - (*Dot Dash*) is the code for the letter *A*. The pattern - • (*Dash Dot*) is the code for the letter *N*. Substitute 0s for dots and 1s for dashes, and Morse code looks an awful lot like binary code: 01 for *Dot Dash*, and 10 for *Dash Dot*.

In the case of binary codes for colors, the number of colors you can define depends on the number of bits you've assigned to keep track of colors. If you assign just 1 bit to a color, you have only 2 codes to use—0 and 1—and only 2 colors you can define—usually black and white, with no grays.

If you assign 2 bits, you have 4 codes you can use, which is enough to keep track of 4 colors:

00
01
10
11

If you assign 3 bits, you have 8 possible codes to let you track 8 colors:

000	100
001	101
010	110
011	111

And so on.

As you might guess, you don't have to use all the codes available for any given number of bits, but you can if you want to. With each additional bit that you assign to keep track of colors, you double the number of colors you can keep track of. As Table 1-1 shows, this doubling of bits adds up pretty quickly. At 8 bits, the number of colors jumps up to 256, at 16 bits it's more than 65,000, and at 24 bits it's more than 16 million. It happens to be a little clumsy to talk about 65,536 colors or 16,777,216 colors, however. That's why, as you get to higher numbers of colors, both people and programs shift from talking about the number of colors to talking about the number of bits.

Table 1-1   Number of Bits Versus Number of Colors

Number of Bits	Number of Colors
1	2
2	4
3	8
4	16
5	32
6	64
7	128
8	256
9	512
10	1024
11	2048
12	4096
13	8192
14	16,384
15	32,768
16	65,536
17	131,072
18	262,144
19	524,288
20	1,048,576
21	2,097,152
22	4,194,304
23	8,388,608
24	16,777,216

You can use this table to convert between numbers of colors and numbers of bits, or, if you don't have the table handy, you can just remember to start with two colors at one bit, and double the number of colors with each additional bit.

That should rarely be necessary, however. Although printers and products that use color liquid crystal display (LCD) screens—like LCD monitors and Pocket PCs—sometimes offer unusual color depths, the overwhelming majority of products that refer to color depth in bits offer one or both of just two choices: 16 bits and 24 bits. If you see mention of more bits, you can treat it as 24 bits.

Why 24 Bits Equals 48 Bits You'll occasionally see references to color depths that are greater than 24 bits (notably 32 bits, 36 bits, and 48 bits), but if you take a close look at these claims, you'll see that what they're really delivering is 24-bit color in almost every case. The additional bits are either used for something else or aren't being used at all.

Some video cards, for example, offer a 32-bit mode, either instead of or in addition to their 24-bit mode. In either case, however, they're using only 24 bits for color. For those video cards that offer both, you won't see any difference in color if you switch Windows to use one mode or the other for the simple reason that there won't be any difference. The 32-bit mode is still using 24-bit color.

Most of the other products that claim to use more than 24 bits are scanners. In some cases, the claim is pure marketing hype because the extra bits don't carry any useful information. Even for those scanners that can take advantage of the additional bits, however, the information is used internally only, in ways that there's no reason to go into here. The point is that what gets handed over to the computer still uses only 24 bits for color.

Why 24 Bits Is a Magic Number

It's important to know that there's a good reason why these color depths are the most common choices.

First, the more bits that your camera, computer, printer, or video card has to shuffle around, the more work it has to do and the longer it takes. More bits also means taking up more space in your camera and on your hard disk. So unless you like waiting for your camera and computer to catch up to you, and you like running out of room, you don't want to use more bits in an image than you need.

The question then becomes what is the smallest number of bits you can get away with. Ask a color scientist, and he or she will tell you that people can see something in the very rough range of 8 million distinct colors. If you take a look at Table 1-1, you'll see that 23 bits would give you a close fit between the number of colors you can see and the number of codes available with 23 bits.

Unfortunately, assigning the 8 million or so codes to just the right colors would not only be extremely time consuming, it would be tricky beyond belief. You'd have to assign the colors carefully, because the human eye is more sensitive to changes in some colors than in others, and because the number of colors you can see varies from one person to the next.

Note   Like most other things about eyesight, you lose the ability to distinguish colors as you get older, so the older you are, the fewer colors you can see.

Jumping up to the next level, at 24 bits, solves that problem. By tracking twice as many colors as you actually need, you don't have to worry about which colors the eye is more sensitive to or whether some people can make fine distinctions that others can't see. Just assign the codes so the colors are spread out evenly, and you'll automatically cover all the bases. This doesn't mean that you can't wind up with poor color using 24-bits, just that it's hard to mess up.

In general, color in photos looks good enough at this level for 24-bit color to have earned the name *true color*, which you can define as being a match for film photography, or, in context of this discussion, as the level of digital color that won't show any steps. And that's why 24 bits is one of the most common choices for color depth.

Lingo   *True color* is the color depth needed so a photo won't show any steps in areas that should shade smoothly. In practice, this means 24-bit color.

Alas, not all computers—particularly older computers—are fast enough to run at tolerable speeds using 24-bit color. The solution is to pick another level of color depth that speeds things up significantly and still gives pretty good color.

As it happens, 16-bit color fills the bill. It leaves your system with only two-thirds as many bits to move around, and offers a level of color called *high color*, which means it has enough colors so a fairly large percentage of photos won't show any steps where they shouldn't.

Lingo   *High color* is the color depth needed so a high percentage of photos won't show steps in areas that should shade smoothly. In practice this means at least 16-bit color, but less than 24-bit color.

Each jump in color depth beyond 16 bits cuts down the number of photos showing steps by even more, and all of these levels also qualify as high color. However, with each jump, you get a smaller and smaller percentage of photos added to the "looks good" pile. As a practical matter, people seem to be satisfied with 16-bit color as a compromise, and the entire computer industry has gone along, which is why high color usually means 16-bit color. To avoid confusion, however, keep in mind that the term really applies to any color depth that uses at least 16 bits but less than 24 bits.

Colors and Shades of Gray

We're finally in a position to answer the question we set out to tackle: how do you translate between number of colors and number of shades of gray? The answer lies in how you combine the shades to make colors.

Consider a single set of red, green, and blue phosphor dots on a monitor, as shown in Figure 1-24.

Figure 1-24 A set of phosphor dots.

Now, assume you have only two shades of gray available for each color dot: full on (which we'll show as filled in) or full off (which we'll show as empty). That gives you eight possible combinations of the three dots, as shown in Figure 1-25.

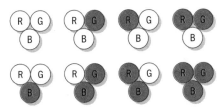

Figure 1-25 All the possible combinations of phosphors if you allow just two levels of gray.

Your eye will interpret each of these eight combinations as a different color. So with two shades of gray, you get eight colors.

If we move up to three shades, you get 27 combinations, which is too much to absorb visually if we show it to you all at once. Figure 1-26 shows just the 9 combinations you get with one shade of red. You'll get another 9 combinations with the second shade of red, and still another 9 with the third shade.

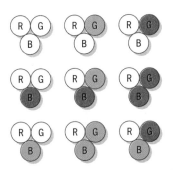

Figure 1-26 This shows one-third of the possible combinations with three levels of gray.

Here too your eye will interpret each of these 27 combinations as a different color. So with three shades of gray, you get 27 colors.

The number of colors that you get climbs quickly when you add a shade of gray, because you're adding it to each of three phosphor dot colors. To get the number of colors, multiply the number of red shades times the number of green shades times the number of blue shades. Table 1-2 shows some of the stops along the way, starting with two, three, and four shades.

Table 1-2   **Number of Shades of Gray Versus Number of Colors**

Number of Shades of Gray	Number of Colors
2	8
3	27
4	64
8	512
16	4096
32	32,768
40	64,000
41	68,921
64	262,144
128	2,097,152
256	16,777,216

There, finally, is the whole story of how to translate between number of colors, number of bits, and number of shades of gray. Also, although you may not have noticed it, there's one additional nugget of interest here. If you compare the last entry in Table 1-1 with the last entry in Table 1-2, you'll see that the number of colors for 24-bit color—16,777,216—is identical to the number of colors for 256 shades of gray. So, 256 shades of gray is 24-bit color is true color. By any name, it's a match for film color.

When Numbers Don't Match   There's one last thing to say about the different ways to measure color depth before we can tie this discussion up in a neat package. If you compare Table 1-1, showing the number of colors for any given number of bits, to Table 1-2, showing the number of colors for any given number of shades of gray, you'll find that for any given number of colors in one table, you won't always find an exact match in the other.

For your convenience in making the comparison, Table 1-3 combines some selected entries from both tables. Take a look in particular at the three highlighted entries hovering around 65,000 colors.

Table 1-3 Shades of Gray, Number of Bits, and Number of Colors

Number of Shades of Gray	Number of Bits	Number of Colors
2	3	8
No match	4	16
No match	8	256
40	No match	64,000
No match	16	65,536
41	No match	68,921
256	24	16,777,216

As you can see in the table, 40 shades of gray define somewhat fewer colors than 16 bits can define, while 41 shades define somewhat more. In case you're wondering, this mismatch doesn't cause a serious problem because the people who design these things are free to ignore some possible color combinations if they don't have enough codes available to cover them all, or to limit themselves to the smaller number of shades and leave some codes undefined. Either way, it just means that the steps between some colors will be larger than the steps between others. With 16-bit color or higher, you won't lose a large percentage of possible colors.

When you're dealing with anything less than 16-bit color, the steps are pretty big in any case. At the lower end of the scale—using 16 colors or 256 colors, for example—the choice of colors to use for each code is basically arbitrary.

How Digital Is Better Than Film and Film Is Better Than Digital

When you come right down to it, digital photography still hasn't caught up to film photography in every way, but it's way ahead of film in others. As we said earlier, the reason most commercial photographers are working with digital photography—and the argument for you to join them—is not that digital photos are better than film or even as good in every way, but that they're good enough. Digital cameras can take pictures with great color and reasonably high resolution, and inkjet printers can turn them into photos that, for the best printers, are indistinguishable from film. However, the process is not the no-brainer that it is with film. On the other hand, if you want to modify your pictures after you take them, digital photography has the upper hand.

In short, film and digital photography each have advantages over the other, and you ought to know what those advantages are. For film, the biggest advantage is quality. For digital photography, it's convenience. Let's start with those.

Why Film Wins on Quality

The most obvious quality advantage that film has over digital photography is
better *resolution*, as measured by the ability to resolve fine detail. This is a different meaning for resolution than when we talk about resolution in pixels,
which we'll refer to as *pixel resolution* when we need to draw a distinction
between the two.

Lingo   *Resolution* has several definitions. One of the more important is resolution as a measure
of the ability to resolve details. Another is a measure of the number of pixels in an image, which
we'll call *pixel resolution*.

Consider the photos in Figure 1-27, for example. Both have the same number of pixels, but the one on the left is noticeably crisper than the one on the
right. In other words, the two pictures have the same resolution as measured in
pixels, but the one on the left has better resolution in the sense that it can
resolve more detail than the one on the right. The difference between these two
kinds of resolution is worth a closer look.

Figure 1-27   These two photos have the same pixel resolution, but the one on the left can resolve detail
better.

Pixel Resolution Versus the Ability to Resolve Detail

The pixel resolution for a digital photo tells you how many pixels are in the
photo. And since a pixel is the smallest unit of the picture, you obviously can't
resolve detail that's smaller than a single pixel. That means the pixel defines a
limit for how good the resolution can be, but it doesn't guarantee that the resolution will, in fact, be as good as that limit. The example in Figure 1-27 shows
the difference. The lack of resolution in the photo on the right makes it look
blurry compared to the one on the left, even though both photos have the same
number of pixels.

Two images with the same number of pixels can differ in their ability to resolve detail for any number of reasons. For example, the resolution for the camera that took one of the pictures could be two, three, four, or more pixels wide, compared to less than one pixel wide for the other. There are also ways to edit the picture to add more pixels. This lets you remove pixelization from a photo, but it won't improve the level of detail you can see. The picture will just be blurry instead of pixelized.

The photo on the left in Figure 1-28, for example, shows the same image as the photo on the right. In fact, we created the photo on the right by starting out with the photo on the left. The version on the left has few enough pixels to show pixelization. You can't see pixels in the photo on the right, because we edited it to have a higher resolution in pixels, but you can't resolve any more detail in the photo either. Instead of pixels, you see a blurred image.

Figure 1-28   We edited the pixelized photo on the left to eliminate pixelization.

Resolution and Film

The resolution advantage for film photography compared to a consumer-level digital camera can be quite dramatic. Figure 1-29 compares an enlargement from a scanned film slide to an equivalent enlargement of a 2-megapixel photo of the same scene. In both cases, the picture is focused on a detail from the original photo—a padlock and door handle. We scanned the slide film photo at 2400 pixels per inch. If we had had a scanner handy with a higher scan resolution, the difference would be even more obvious.

Figure 1-29   Film resolution compared to digital resolution.

As you can see, it's easy to make out the padlock and handle in the top photo (taken from a slide). You can also see the wood grain in the log cabin exterior to the right of the door. In the digital version on the bottom, the padlock is hard to make out as a padlock at this level of enlargement, and the wood grain is hidden by artifacts in the image.

Film cameras have better resolution than digital cameras for several reasons. The pixel count for today's digital cameras is one of those reasons. Even the most expensive digital cameras aimed at professional photographers don't offer enough pixels to fully match film. Consumer-level digital cameras offer far less. As a general rule, today's digital cameras offer good enough resolution to let you print a picture at 8 × 10 inches, but if you need to crop out part of the picture before you enlarge it, or you need something bigger than 8 × 10, you may begin to notice the lack of resolution, depending on how much you enlarge the photo.

Film cameras also tend to have better lenses than similarly priced digital cameras. It's not just the quality of the optics that matters, but the range of features available—things like the level of zoom, the ability to change lenses quickly and easily, and whether you have a wide range of lenses to choose from. Each of these advantages for film cameras translates to a limitation for digital cameras. More important, each limitation can force you to use a workaround that results in a lower resolution image than you would wind up with if you didn't need a workaround.

For example, the zoom features in consumer-level digital cameras often won't let you zoom in to the levels of magnification you would ideally want. That means you often have to take a wider angle shot than you want, crop it on your computer, and enlarge it to get the shot you wanted in the first place. Alas, enlarging the image is precisely what will make you run into resolution limitations.

There are also other issues that affect picture quality besides the camera itself. If you want printed pictures with a digital camera, you need to print them. There are ways to have that done professionally—just like you drop off film at a developing lab—but the more obvious, more immediate way is to print them yourself. As we mentioned earlier, the best inkjet printers can print photos that are indistinguishable from film prints. Unfortunately, most inkjet printers offer somewhat lower output quality, and, adding insult to injury, photos printed by inkjets tend to fade or discolor much more quickly than film prints as well.

With digital photos there is also an occasional issue with printing colors that are a reasonable match for what the camera sees. In years past, matching colors between camera, monitor, and printer was a major headache. It's gotten a lot easier today, but it still crops up as a problem from time to time, as we'll discuss in Chapter 5, "Special Issues for Digital Photography."

Color matching issues are inherent in the nature of color, and they create problems for everything from film prints to matching colors in rolls of wallpaper. However, film has the advantage over digital photography of having been around for long enough that the film industry has learned how to deal with color matching in film. More important, to the extent that it's still a problem, it's not something you have to worry about. Just drop the film off for developing, and otherwise ignore it. Some technician will take care of making sure the colors look right. When you print the photos yourself, you're the only technician in sight.

Having said all this, we should also point out that the quality advantage isn't all in favor of film. Digital photos have some quality advantages too. To begin with, digital photos don't deteriorate the way film negatives do. Inkjet prints may not last as long as a photographic print, but if you store the images on a long-term storage medium, like a recordable CD (CD-R), you'll be able to reproduce a brand new, fresh print any time you need one, well into the foreseeable future.

Digital photos can also have better color fidelity than film—despite the occasional problems with color matching. If you use the wrong film, it will tend to bring out the blue in shadows in daylight, or orange in incandescent light. Digital cameras don't shift colors with changes in lighting because they adjust to the light.

Finally, and most important, the quality level for digital cameras is plenty good enough for most purposes. Color quality for many inexpensive digital cameras is excellent; so is the ability to hold details across the range from the darkest to lightest areas in a photo. If all you want to do is put a picture on the Web, or take a snapshot, or otherwise use pictures in a way that doesn't require a great deal of enlargement, you won't run into any resolution issues. In those cases, your photos can look just as good as anything you can take with film. Maybe better.

Why Digital Photography Wins on Convenience

Some people would argue with the title of this section, and insist that film is much more convenient than digital photography. What they mean is that with film all you have to do is take the pictures, drop them off at a one-hour developing lab, go shopping, and then stop by later to pick up the prints.

If that's all you want to do with your photos, then they're right; film is more convenient. But if you want to do anything else, the balance tips toward digital photography.

If you want to e-mail your photos, put them on the Web, edit them, or print your own copies, digital photography is far more convenient. The instant you take your photo, it's available in the format you need. Simply move the image over to your computer (which can be as easy as plugging in a cable between the camera and computer) and there it is, ready to go. If you want to do the same with film, you first have to develop the film, scan it into your computer, and save it as a file. If you need to scan an entire roll's worth of pictures, this can quickly get tiresome.

There's also a tremendous convenience in having the photos available as soon as you take them. Polaroid built a company on being able to develop pictures in 60 seconds. With most digital cameras, you can look at the pictures on a built-in LCD as soon as you snap the picture. And if you want to send a picture to Grandma, you can snap the picture, plug in the cable, and send it as an e-mail attachment while the Polaroid camera's shot is still developing.

Digital photos are also much more convenient to store than film negatives. Storing files takes up a lot less space than storing physical negatives, and the files are easier to search through to find the right picture. Storing digital photos can also be remarkably inexpensive, thanks to CD-R discs.

Oh, and one other thing: you don't have to drive to the photo lab to drop off the film.

Other Advantages: A Point-by-Point Comparison

In addition to the major advantages we've just mentioned for both film and digital photography, there are any number of other advantages for one or the other. Most of these are straightforward and don't need much of an explanation. Here then are what amount to two glorified lists: the advantages of film and the advantages of digital photography. We'll start with film.

Advantages for Film over Digital Photography

We're really not trying to stack the deck, but the advantages film has over digital photography (other than better resolution) make for a short list:

■ For any given level of features, including the availability of accessories like additional lenses, film cameras are much less expensive than equivalent digital cameras. However, you have to balance the savings on the camera with the fact that you have to pay for film and for processing all your shots—including the ones you took with your fingers over the lens or otherwise don't want to print.

■ You don't have to worry about your batteries running down or carrying extra sets of batteries with you, the way you do with a digital camera.

■ Carrying extra film with you is a lot easier than having to find some way to offload photos from a digital camera that doesn't have room for any more. It's also a lot cheaper than buying additional storage cards for your digital camera.

Advantages for Digital Photography over Film

In contrast to the short list of advantages for film, digital photography has lots of advantages over film:

- Digital cameras may be more expensive to buy than film cameras, but using them is far less expensive. Film costs money, whether you like the picture or not. With digital cameras you don't have to print or pay for the pictures you don't like. Similarly, there are times when you need to take a picture for a specific purpose, like providing one for a conference you need to attend. With digital photography, you'll never have to take additional pictures you don't want just so you can finish up the roll of film and develop the picture you need. If you don't like your digital photo when you see it on the camera's LCD you can delete it, or you can look at it later on your computer screen and decide to delete it. Either way, you recover the space with no cost involved. Even for the pictures you keep, you only have to pay for printing the ones that you want hard copies of—which is a lot less expensive than developing an entire roll of film. And you never have to pay to buy new film either. Just delete everything in your camera storage, and start over again. You have to replace batteries far more often than with a film camera, but you can use rechargeable batteries to minimize the cost.

- Because there's no film, you don't have to worry about storing film properly if you keep it around before using it, and you don't have to worry about film's limited shelf life.

- Because digital photos are virtually free (if you don't count the cost of the batteries) you can take as many as you want. Go ahead and take 300 shots of your daughter's college graduation; it won't cost you anything. Just delete the ones you don't like, store any that you're not sure of, and print only the ones you really want.

- Not only do most digital cameras let you see a picture as soon as you take it, thanks to a built-in LCD, but many offer a video output port so you can see the pictures on your TV.

- Many digital cameras include extra features, like letting you shoot short videos, or add sound notes to your photos.

- Editing a photo or retouching a photo for problems like red eye is trivial with digital photography. With film, you'd have to get a print of the photo, then scan it into your computer before you could fix the problem. And you'd have to buy the software somewhere. With a digital camera, the photo is already in digital form, and your camera probably came with the software you need.

■ You don't have to worry about negatives or slides getting scratched, dirty, or dusty, or worry about them fading over time.

Digital Photography Versus Film for What You Want to Do

All of these comparisons of digital photography to film are fine as far as they go, but what you really need to think about is what you want to do with the camera and how these issues apply to the kinds of pictures you want to take. These are questions you'll have to answer for yourself, but we can give you some suggestions.

If you're looking to take pictures with the idea of cropping in on small details and enlarging the images, film is certainly the way to go. On the other hand, if you don't expect to do much cropping and enlarging, the benefits of film will be minimal, and the argument for staying with film is harder to make. The only reason to stay with film in that case is if you're absolutely sure you'll never want to put a picture on the Web, e-mail one as an attachment, print a picture yourself, or edit an image to enhance it or remove flaws.

If you need the ability to enlarge photos to the point where film's better resolution is a necessity, but you also want to have one or more of these digital capabilities, you might want to use both a digital camera and film camera. Alternatively, you might want to stay with film, but get a scanner so you can convert your pictures to digital format when you need to.

If the kinds of photography you're interested in don't include the need to enlarge images very much, or very often, a digital camera may be your preferred choice, even if all you want to do is take snapshots of friends and family. If you also want the convenience of being able to easily e-mail a photo or put it on the Web, or you want the ability to enhance your photos and then print them yourself, or any combination of these capabilities, digital photography is certainly the way to go.

Key Points

■ Digital images aren't better than images on film, or even as good in every way, but they are good enough for most purposes.

■ The essential difference between digital and analog technologies is how they make changes—in color, brightness, audio volume, or some other factor. Analog technologies change in a smooth, continuous way. Digital technologies break changes into discrete steps.

- If a digital technology takes small enough steps—such as stepping through the shades of blue in a picture of the sky—the result will be indistinguishable from a continuous, analog technology.

- Digital photos are built from discrete pixels. One of the most significant improvements in digital cameras has been the growth in pixel count at increasingly affordable prices.

- All the colors you can see can be reproduced by combining just three colors. For digital cameras, these primary colors are red, green, and blue.

- Talking about shades of gray is convenient shorthand for talking about shades of each primary color. 256 shades of gray translates to 256 shades of red, 256 shades of green, and 256 shades of blue.

- What's true for a monochrome image with a particular number of shades of gray is also true for a color image with the same number of shades of gray.

- True color—a level that makes color in digital photos indistinguishable from film—is 24-bit color, which is the same thing as 256 shades of gray, or roughly 16.7 million colors.

- The biggest advantage film has over digital photography is quality. The biggest advantage that digital photography has over film is convenience.

Chapter 2

Knowing (and Choosing) Your Camera

This chapter is for two groups of people: those who have already bought a digital camera, and those who haven't.

First and foremost, if you already have a camera and you're still learning your way around its features, this chapter will help you get familiar with your camera's capabilities and what it can and can't do. (We won't get into the particulars of how to take advantage of any features here, however. You'll find those details in Chapter 3, "Getting Started with Digital Photography.") As it happens, the same information can serve as a buying guide.

Choosing the right camera is a lot like choosing the right breed of dog to bring into your life. There are lots of choices out there. Each breed has its own distinct personality, a particular size and weight, and a need for a particular level of exercise. You just have to find the one that matches your lifestyle. If you want to take short walks, you don't want to take a Siberian husky home with you only to find out that the reason huskies make great sled dogs is that they love to walk and need the exercise.

There are lots of choices with cameras too, and just as with picking the right breed of dog, the trick with cameras is to find the right fit. If you haven't bought a camera yet, this chapter can help you sort out the choices and hopefully make it a little easier to pick the right one.

If you've already bought a camera and you're happy with the result, the chapter may help you think about the categories of cameras in a different and more useful way. It may also bring up some issues that you'll want to consider for your next camera.

If you've just bought a camera recently, and your research before buying was limited to, say, spotting a low price at a consumer electronics store and buying the camera on the spot, this chapter's *really* for you. You may want some reassurance after the fact that you made a good choice (especially if you still have the option of returning the camera for a full refund within some number of days). This chapter will help you find out whether you inadvertently brought home a husky, even though you hate to walk.

Categories of Cameras

Digital cameras come in as many varieties as film cameras, if not more. They range from basic point-and-shoot cameras, with few or no settings available, to cameras appropriate to control freaks who like to manage all the details. Some are aimed at rank beginners, others at professional photographers, and still others at every level in between. Some offer minimal resolution, by which we mean a low pixel count; others offer lots of resolution.

That's just the beginning. You can find digital cameras that are limited to using the built-in lens they came with and others that can use additional lenses that are available as accessories; cameras with zoom and cameras without zoom; cameras that come with lots of bundled software, and cameras that come with little or no software; and...well, you get the idea. If you can think of a variation on a feature, you can probably find a digital camera that offers it.

Having plenty of choices is great, until you try to sort everything out. At that point, it helps to think about categories as a way to understand your camera in context, if you already have one, or figure out which category you should be looking at to choose a camera if you don't already have one. Unfortunately, that's hard to do with digital cameras, because there isn't a single scale, or even just two or three scales, you can measure cameras against to categorize and compare them. Each camera model has its own nearly unique mix of features, target users, and capabilities. And that complicates matters.

Choosing a camera is what you might call a multidimensional problem. Instead of having a single dimension with categories that run from beginner to advanced, or low resolution to high resolution, or poor lenses to great lenses, you have a whole bunch of dimensions—including the three we just mentioned— each with its own set of categories. That means you need to look at each set of categories separately, to see which category the camera falls into in each dimension.

Looking at one set of categories at a time can help you better understand a camera's features, especially in relation to other cameras. It can also help you pick a camera if you don't have one already. Start with the feature you consider most important—which we'd argue is resolution—and eliminate any cameras that aren't in the right category for that feature. Then go through the rest of the features, starting with the ones you consider most important, crossing cameras off your list as you go. Keep in mind, however, that what counts as important depends very much on what kind of photographer you are and what kinds of pictures you take, which are the two issues we'll start with.

What Kind of Photographer Are You?

Start by deciding what kind of photographer you are. You basically have three choices. For lack of better terms, we'll call them point-and-shoot, mildly creative, and prosumer. (We're ignoring professionals in this book.)

If you're a point-and-shoot photographer, you don't want to learn any more about photography than you have to. You don't care if you have control over shutter speed or depth of field, and you probably don't even care what depth of field means. (We'll be discussing these sorts of features later in this chapter. You can skip over them if you don't care about them, but we recommend that you at least skim through them so you have some idea of the possibilities.) Most of all, you don't have any great need to change settings and don't particularly want to spend the time learning what settings to change. All you want to do is pick up a camera, look through the viewfinder or at the liquid crystal display (LCD), and click the shutter button. You probably don't mind having additional capabilities available that you might, conceivably, explore one day—as long as they don't get in the way and confuse you today. But you'll probably never use them. What you absolutely need is an easy-to-use, fully automatic mode.

If you're a mildly creative photographer, you're a little more adventuresome. You probably want to use a fully automatic mode more often than not, but you also want some additional features. You might want a *macro mode*, for example, which lets you take pictures from distances measured in inches—as with the close-up of a fern shown in Figure 2-1—instead of the standard camera minimum of several feet. You certainly want the ability to zoom. You would probably enjoy exploring other capabilities of the camera as well, but you can easily live without many of them. You would probably not be happy with a camera that offered nothing beyond point-and-shoot capability.

Lingo   *Macro mode* lets you take photos from distances as close as an inch or two, far closer than the standard for most lenses.

Figure 2-1   This close-up of a fern required macro mode.

If you're a prosumer, a consumer whose knowledge of and experience with photography approaches a professional photographer's level, you want it all—or at least as much as you can afford. You probably want a macro mode, a better zoom than most digital cameras offer, a choice of modes that runs from full automatic to full manual with all stops in between, accessories that include an assortment of lenses, and more. You also want the highest resolution you can afford, because you expect to crop your pictures and enlarge them. Unfortunately, unless you have lots of money to spare, you won't be able to get a camera with all the features you probably want. The good news is that you can get most of the features you want at a reasonable price.

What Kind of Photographs Do You Want?

The photographs you plan to take are related to the kind of photographer you are, but you should consider the specific kinds of photos you expect to take. We've listed a range of possibilities to help you get started and give you some ideas. You might want to write out a list of the types of photographs you generally take and want to take, and then look at what features you need in order to take them.

Medium-Range Snapshots

Medium-range snapshots are the kinds of pictures most people take most often—the shot of one to five people grouped together at a wedding, around the pool at a barbecue, or standing in front of the Eiffel Tower to prove they were there. The distance between the camera and the subject is typically 5 to 10 feet, as in the two photos shown in Figure 2-2.

Figure 2-2   Most people take medium-range shots like these more often than they take any other kind of picture.

Any camera should be able to handle this kind of photograph; to the extent that there is a standard photograph, this is it. Look in most family albums and you'll find that something over 90 percent of the photos are taken at medium range.

Close-Ups

As a rule, the closer you get to your subject, the better the picture is likely to look, whether your subject is a person's face, your kitten, or your new car. One benefit is that you get to see more of the subject. This is especially true with photos of people. The eyes may or may not be the window to the soul, but if you can see the eyes clearly, it makes for a better photo. Another reason to get in close is to cut out all the extraneous stuff that is going on around the subject and focus the viewer's attention on what you meant to take a picture of in the first place. You can see the difference a close-up makes by comparing the two photos in Figure 2-3.

Figure 2-3 Getting close in focuses the viewer's attention on the subject.

To get this sort of close-up of a person's face, you need to be able to focus as close as 3 feet. Note that some cameras can't focus any closer than 3 to 4 feet, which doesn't give you much room to work with.

Indoor Shot

In the bad old days, taking a good photograph indoors required either some deep knowledge of photography or lots of luck. Good lighting indoors was scarce, flash was hard to use properly, and the odds of getting a good photograph without knowing what you were doing were slim. Today, of course, almost all cameras—digital included—have some kind of built-in flash, an external flash attachment, or both. More important, they have automatic flash modes that take care of deciding when to use flash to light up the scene.

Digital cameras have the additional advantage of being more sensitive than film, so you can take pictures indoors even without using flash and without needing much light. Digital cameras have another advantage also. As you may know, colors actually change with changes in lighting—between sunlight and incandescent light, for example. With film cameras, you have to match the film to the lighting to get good color. With digital cameras, you don't have to worry about this issue.

Digital cameras use a feature called *white balance*, which adjusts the colors the camera sees to match the current lighting conditions. More precisely, the camera adjusts the balance of red, green, and blue to make sure white appears white. Other colors fall into place. (Some cameras don't do as good a job at this as others, but that's a separate issue.)

Lingo   Colors change with different lighting conditions. Digital cameras maintain good quality color in different lighting conditions by adjusting the *white balance*—the balance of red, green, and blue—to make sure white appears white.

This combination of features makes it easy to switch between taking pictures outdoors and indoors, and also take pictures indoors with or without flash, like the ones shown in Figure 2-4. The photo on the top uses ambient incandescent light only; the photo on the bottom uses the camera's built-in flash.

Figure 2-4   Digital cameras can use ambient light indoors or use flash.

Although most digital cameras have a built-in flash suitable for taking indoor shots from a few feet away, the flash will rarely put out enough light to take high-quality pictures at longer distances to the subject. If you need a powerful enough flash to fill a room, you'll need a hot shoe attachment, which lets you add a (more powerful) external flash.

Closing in from a Distance

Close-ups aren't always the result of getting close to your subject. Sometimes they come from staying well back. Consider the close-up on the left in Figure 2-5 compared to the one on the right, for example. Unlike the photo on the left, the photo on the right is a candid shot, courtesy of a telephoto lens.

Figure 2-5   You can get close-ups by getting in close, which we did for the photo on the left, or by using a telephoto lens, as in the photo on the right.

For things like mountains on the horizon, being able to get close up from far away will save you a lot of walking. For things like taking a shot of your favorite ballplayer during a game, it will let you get close up without getting arrested for going onto the field. For photographing people in general, it will let you get spontaneous shots from a distance without distracting your subject or having him or her start posing for you. (Some people, like the woman in the photo on the left, who is obviously aware of the camera, can pose in a natural, relaxed way. Others, however, can't pose without looking stiff, which is the best way we know to ruin an otherwise good shot.) For pets and other animals, a telephoto lens can likewise let you get a spontaneous shot without distractions. Pets don't pose, but they have an annoying habit of stopping whatever they were doing that made you grab the camera and getting interested in the camera instead (toddlers tend to do the same thing).

Don't confuse telephoto lenses with zoom lenses, which aren't necessarily telephoto. In some cases a zoom lens zooms from wide angle to normal view. To get telephoto capability, you may need to buy a separate lens. If so, and if you want this capability, you'll need to make sure not only that the camera has some way to attach lenses, but that there are actually lenses available for the camera.

Extreme Close-Ups

By extreme close-ups we mean close enough to see the dew on a flower, as in Figure 2-6. This kind of photography is a lot of fun because you get to see your subject in a new and fascinating way, up close and personal.

Figure 2-6   Some cameras' macro modes let you get close enough to see dew on a flower.

As we mentioned earlier, this sort of extreme close-up requires a macro mode, but not all macro modes are created equal. Some let you take pictures as close as an inch or two from the subject. Others limit you to a closest distance of four or five inches.

Photos That Need Special Lenses or Filters

Telephoto lenses aren't the only kind of lens that you might need. A wide angle lens can help widen your field of view, as shown in Figure 2-7. They'll let you photograph a building, the inside of a room, or a wider swath of an outdoor scene even when you can't move further back to get the whole scene in view with the camera's standard lens.

Figure 2-7   Wide angle lenses let you photograph more of a scene without stepping back.

Wide angle lenses can actually go all the way up to seeing a full 180 degrees, using what's called a *fisheye lens*. The result looks distorted, but visually interesting, as you can see in Figure 2-8.

Figure 2-8   As a category, wide angle lenses go all the way to a 180-degree view for a fisheye lens.

As with a telephoto lens, if you are interested in taking any of these kinds of shots, you'll need to confirm that the camera has some way to attach lenses and that there are actually appropriate lenses available for the camera.

Panoramas

Sometimes, you'll see a glorious panoramic view that you just have to capture—like a mountain on the horizon reaching up to the sky, a seashore seen from a boat, or simply an attractive landscape, as in Figure 2-9.

Figure 2-9 This shot doesn't show off the view very well.

As you can see in the figure, pictures like this tend to be disappointing if you use a standard format or *aspect ratio*, the ratio between the width and height of the picture. (More precisely, the aspect ratio is the ratio between the larger dimension of the picture and the smaller dimension, which keeps it the same whether the picture is in portrait or landscape orientation. The aspect ratio for 35mm film, for example, is 1.5 to 1, usually written as 1.5:1.) With a standard aspect ratio, the landscape that looked so impressive in real life gets lost in the image.

Lingo   The *aspect ratio* of a photo (or video screen) is the ratio between the larger dimension of the rectangle and the smaller dimension.

To make this sort of picture work visually, you need to change the aspect ratio to make the area you want to highlight stand out. This is a variation on the rule that the closer you get to your subject the better the picture will look. The point is to get rid of all the extraneous parts of the picture that can draw your attention away from the part you should focus on. Figure 2-10 shows what a shot of the horizon can look like if you change the aspect ratio.

Figure 2-10   Change the aspect ratio, and you can improve the shot.

You can get from the picture in Figure 2-9 to the one in Figure 2-10 by cropping the first picture. With digital cameras that's easy to do on your computer, of course. Until you get to your computer, however, the extra pixels that you're going to get rid of later use up space in your camera's storage, and could keep you from taking another shot because you don't have room for it. If you tend to take a lot of this kind of picture, you'll ideally want a camera with a panorama mode, which automatically crops the image to change the aspect ratio and save on storage space. However, don't confuse this kind of panorama mode with the stitched panorama mode that we cover next.

Stitched Panoramas

Digital cameras can let you do some things relatively easily that film cameras barely let you do at all (at least, not without scanning the photo to turn the film image into a digital image). One of those things is to take multiple shots for a panorama so you can stitch the images together for a truly expansive view, like the one shown in Figure 2-11.

Figure 2-11   This panorama stitches four shots together to give a wide view.

We'll discuss stitching pictures together in detail in the second part of this book. For the moment, it's enough to know that it can be done. To give you a sense of what's involved, however, Figure 2-12 shows the individual photos that we stitched together to create Figure 2-11. Notice that each of the pictures overlaps a bit with the one next to it.

Figure 2-12 These are the individual pictures, before stitching them together.

Any digital camera will let you take photos that you can stitch together; stitching is something that you do with software and has nothing to do with the camera as such. If you plan to do this sort of thing very often, however, you'll find it useful to have a camera that has a stitching panoramic mode. The best of these will help you aim the camera so the shots overlap properly. When in panorama mode, their LCDs will show you a ghost image of the edge of the previous shot, so you can line up each shot properly with the one before.

Note   If you plan to take multiple shots to stitch together, also plan on buying a tripod, and make sure the camera can attach to a tripod.

Rapid-Fire Stills for Capturing Action

Sometimes, you know something is going to happen, but you don't know quite when—like the teenager starting to head for a cannonball in the pool in the upper left photo in Figure 2-13. If you can shoot pictures every half-second or so, as with the photos in Figure 2-13, you're not guaranteed to get the perfect shot right on the money, but you have a good chance of getting a good shot, which in this case would be on the bottom left, with the subject of your shot in the air, about to hit the water.

Figure 2-13 Continuous shooting gives you a good chance of capturing the action.

If you expect to take this sort of photo, you'll want a camera that can take multiple shots automatically without you doing anything but holding the shutter button down. The less time the camera needs between shots, the better. This feature goes under different names, including *burst mode* and *continuous shooting*.

Lingo *Burst mode* and *continuous shooting* are two names for a feature that automatically shoots multiple pictures as you hold the shutter button down.

Full-Motion Video

You can't use a digital camera to replace a video recorder, but many digital cameras will record video nevertheless. The sequence of pictures in Figure 2-14 was taken from a 25-second video that we shot with a digital camera at 15 frames per second and 320 × 240 pixel resolution. If you don't have a video camera—or even if you do have a video camera, but you don't have a *digital* video camera—you may find this feature useful. If you've just moved to a new house, for example, you can use a video to pan across your house and property to show it off to your friends, instead of sending a single, still shot.

Figure 2-14   These are individual frames from a 25-second video shot with a digital camera.

Not all digital cameras offer this feature. If it's something you want, you'll have to make sure it's available in the camera you choose.

How Easy Is the Camera to Carry?

It's a well-known rule in politics that you can't do anything unless you get elected. There's a similar rule in photography: you can't take a picture unless you have your camera with you. If you're a decidedly casual photographer who only takes your camera out for birthdays, weddings, and other special occasions, that's not an issue. But if you want to take pictures more often, and more spontaneously, not having your camera with you can be a big problem.

Cameras range in size all the way from big and bulky to small enough to fit comfortably in a shirt pocket. You can even find cameras that are literally the size of a small stack of credit cards, as you can see in Figure 2-15. One of the questions you have to answer for yourself is how small and how portable a camera you need. That can be a particularly tricky issue if you're looking to buy your one and only camera, since you may have to choose between easy portability and other features you'd like to have.

If you're a point-and-shoot photographer and you don't want to be confused with a lot of features in any case, this is an easy choice. Go for portability. For mildly creative photographers and prosumers, however, this is a tough choice—and not one that we can give much help with. You'll have to weigh the relative merits of lots of features on the one hand versus easy portability on the other, and decide what balance you're willing to live with. It may be worth it to you to do without a macro feature or the ability to add a different lens, for example, in exchange for being able to take the camera wherever you go. Or it may not be. Make sure you give it enough thought to make the right decision.

Figure 2-15   Some cameras are as small as (but not quite as thin as) a credit card.

If you're serious enough about photography that you have, or are planning to get, a second camera, you have much more wiggle room. In that case, you might want to think of one camera as your primary, or best, camera. The other will be a secondary camera that you can take with you anywhere, any time without a second thought.

Feel free to go crazy for features on the primary camera and not worry much about portability. This is the camera you'll use when you're seriously out hunting for photos with premeditation. But feel just as free to sacrifice features for portability on the secondary camera. Granted, you may miss some shots because your secondary camera doesn't have some feature you'd like, but you'll have a camera with you, which means you can get at least some shots that you would otherwise surely miss.

Choose a Resolution: How Much Do You Need?

As we suggested earlier, the most critical choice you can make in choosing a camera is choosing the resolution. More resolution is always better, assuming it's within your budget, but the real question is how much you need.

There are two factors that determine the answer: whether you will ever enlarge pictures and whether you will print them. If you never crop a picture and then enlarge it, and never print a picture larger than 5 × 7 inches, you don't need a camera with much resolution at all. If you plan to do either of these—to enlarge the picture or print larger than 5 × 7 inches—you need a bit more. If you plan to do both, you need a lot more. Table 2-1 shows the minimum resolution you should be considering depending on what you want to do.

Table 2-1   **Minimum Camera Resolution for Different Uses**

Use	View on Screen Only	Print at 5 × 7 Inches	Print at 8 × 10 Inches
Will never enlarge	1 megapixel	1 megapixel	2 megapixels
Will sometimes enlarge up to four times the size	2 megapixels	3 megapixels	7 megapixels

Here's a warning (or maybe what lawyers would call a disclaimer): if you're a point-and-shoot photographer, you've probably never cropped a picture and rarely printed one at 8 × 10 inches. Now that you have your hands on (or are about to get your hands on) a digital camera, that's likely to change. One of the great things about digital cameras is that after you move the photos to your computer, you can do all sorts of things with them far more easily than you can with film, including cropping them and printing them on your printer. (We'll cover all that in later chapters.) So just because you've never done these things before, don't assume you never will.

We also have to stress that these are *minimum* megapixel ratings. You may be pickier than we're being here, or have some special requirement that needs a higher resolution. With that in mind, it will be helpful for you to know how we came up with these numbers. Then you can decide whether you want to argue with them.

What the Minimum Resolutions Are Based On

If you're viewing pictures on screen only, you never need more pixels in the photo than there are pixels on the screen. So the number of pixels you need is defined by the screen resolution.

We'll go into the specifics in more detail in Chapter 3 when we discuss the camera resolutions to use for each screen resolution. For the moment, it's enough to know that a 1-megapixel camera can produce more than enough pixels to fill a screen using 1024 × 768 resolution. Considering that you will usually want some room left over for a program window, title bar, menu, and so on, a 1-megapixel camera should be more than adequate for screen resolutions up to 1280 × 1024, which is more than most people use. Even at 1600 × 1200, it will fill a good portion of the screen without enlarging the image. If you want an image that fills a screen at 1600 × 1200 resolution, however, you'll need a 2-megapixel image instead.

If you're printing, the important number becomes pixels per inch, not pixels for the entire photo. At 7 inches across on the page, a photo that's 1024 pixels across is just under 150 pixels per inch (1024 divided by 7 is 146.3). In the other dimension it's a little over 150 pixels per inch (768 divided by 5 is 153.6). Here again, we'll cover the specifics in Chapter 3, when we discuss camera resolutions to use. What you need to know for the moment is that 150 pixels per inch produces reasonably high-quality photos.

For printing at 8 × 10 inches, a 1-megapixel camera is hopelessly inadequate, and a 2-megapixel camera is arguably acceptable. A 2-megapixel camera can produce at least 1200 × 1600 resolution, which translates to about 160 pixels per inch in the long direction (1600 divided by 10 inches) and 150 pixels per inch in the short direction (1200 divided by 8).

That covers the numbers in the first row of the table. The second row involves enlarging the photo up to four times its size. That's actually ambiguous. Let's clarify it.

If you start with a picture that's, say, 3 inches wide by 2 inches tall, and you double the width to 6 inches, the area for the picture doubles—going from 6 square inches (2 inches times 3 inches) to 12 square inches (2 inches times 6 inches). If you double the height of the picture also, the area jumps to four times the size of the original—going from 6 square inches to 24 square inches (4 inches times 6 inches). That's what we mean by enlarging a photo up to four times its size: doubling the size in each direction.

Here's how this translates into the world of pixels:

■ You take a picture with a 2-megapixel camera at 1600 × 1200 pixels.

■ If you view the image on screen at SVGA resolution (800 × 600 pixels) and set your software to use one pixel on screen for each pixel in the image, you'll be able to see only a quarter of the image at once.

■ Another choice is to shrink the image down to see it all on screen at once. That essentially tells your software to use only one pixel on screen for every four pixels in the image, and any given 800 × 600 pixel area of the image will fill only one quarter of the screen, or 400 × 300 screen pixels.

■ Now crop the picture so it's one fourth the size—at 800 × 600 pixels. If you're still looking at the image in shrunken form it will fill only one quarter of the screen, but the image itself still has 800 × 600 pixels.

■ Next, enlarge the image to full-screen size by setting your software back to show one pixel on screen for each pixel in the image.

That's what we mean by enlarging to four times the size. What's more, this specific example explains how we come to the conclusion that a 2-megapixel camera is the acceptable minimum for viewing pictures on screen with enlargements up to four times the size. Simply put, with a 2-megapixel camera, you can crop and then enlarge up to four times the size and still have an 800 × 600 pixel image.

The logic for printing is similar, but we'll get to it from a different direction. The question to ask here is how many pixels you'll need to have at least 150 pixels per inch for your printed picture, both in width and height. At 5 × 7 inches, the answer is:

- 1050 pixels in width (7 inches times 150 pixels per inch)
- 750 pixels in height (5 inches times 150 pixels per inch)

Multiply 1050 by 750, and you get 787,500, which is the number of pixels you'll need in the photo after you finish cropping it to one-fourth the size. To get there, you have to start out with four times as many pixels, which comes out to 3.15 megapixels. That means you need a camera in the 3-megapixel range.

The same train of logic applies at 8 × 10 inches. Unfortunately, the answer at 8 × 10 inches comes out to 7.2 megapixels as your starting number before you crop anything. The day will certainly come when you can get a 7-megapixel camera or better at a consumer-level price, but it's not even close as we write this. The moral here is that whatever the current prices for cameras as you read this, if you plan to enlarge your photos, get the highest resolution you can afford. And that applies whether you as a photographer fall in the point-and-shoot, mildly creative, or prosumer category.

When Your Camera Has More Pixels Than Your Screen Although the image you use on screen will look best when you show it using one pixel on screen for each pixel in the image, it's not a problem to start with more pixels in the image than you eventually want to see on screen. Briefly, if you have more pixels in an image than you have pixels available on the screen, the image either won't fit completely on screen at once, or, if you shrink the picture to fit on screen, the extra pixels will simply be thrown away.

Shrinking the picture, which essentially tells your computer to use only one screen pixel for every so many pixels in the image, can make the image look worse, but it doesn't have to. If it's done properly—which usually means using a photo editing program to change the number of pixels actually in the photo—you can remove pixels without hurting the image quality. We'll cover that trick in Chapter 8, "Fun with Pictures: Basic Editing." The important point for right now is that if you have more pixels than you need, you can adjust the number downward without hurting image quality.

We should also mention that if you have fewer pixels than you need to fill the screen, you can edit the photo to add more pixels. However, when you add pixels this way, you will not get the same quality image as if the original photo included the right number of pixels. We'll discuss this in detail in Chapter 8.

If You're Picky, You Need More

Now we get to the part where your individual level of pickiness comes in—which is more likely to be an issue if you fall in the mildly creative or prosumer category. You may or may not agree with us that 150 pixels per inch is enough. If you use a 200 pixel per inch photo instead of the same image at 150 pixels per inch, you'll be able to see the difference with most printers. However, the difference is subtle, and most people would have to hold the two pictures next to each other to see it. That said, if you want the extra quality that 200 pixels per inch delivers, the megapixel ratings you need shoot up, as shown in Table 2-2.

Table 2-2   **Preferred Camera Resolution for Printing at 200 Pixels per Inch**

Use	Print at 5 × 7 Inches	Print at 8 × 10 Inches
Will never enlarge	1.4 megapixel	3.2 megapixels
Will sometimes enlarge up to four times the size	5.6 megapixels	12.8 megapixels

As the table shows, at 200 pixels per inch, a 5 × 7 photo needs 1.4 megapixels (1000 pixels × 1400 pixels) and an 8 × 10 photo needs 3.2 megapixels (1600 × 2000 pixels). For enlarging at up to four times the size, the ratings go through the roof, at 5.6 megapixels for a 5 × 7 print and 12.8 megapixels for 8 × 10 inches. At least that's what you would ideally want. In the real world, you'll have to settle, once again, for the highest resolution you can afford.

Choosing a Lens System

In the days before digital cameras, a lot of photographers would have argued that the camera was their least important piece of equipment. The camera, they said, was merely a box for holding film, and, more important, a place to mount a lens. It was the lens—or set of lenses—available for the camera that was far and away the most important weapon in the photographer's arsenal.

Even before digital cameras, people who took this position were being a bit extreme. Today, with digital cameras—and even today's film cameras with all sorts of built-in smarts—the argument is even harder to make. Yet the fact remains that the lens or lenses available for a camera are one of the most important factors determining how good your pictures look.

We talked about this a little bit in Chapter 1, "Everything's Coming up Digital," when we pointed out that film cameras tend to have better lens systems than digital cameras that cost about the same amount. However, there's quite a bit more that you should know about lenses.

Issues for a lens system include the following:

- The quality of the optics
- Whether you can change lenses
- If you can change lenses, what lenses are available and how quickly and easily you can change them
- Whether the camera includes zoom
- If it includes zoom, what range the zoom offers
- Whether the lens allows a macro mode
- If there is a macro mode, how close you can get and still be in focus
- Whether the camera lets you see the picture you're about to take by looking through the same lens as the camera uses to take the shot

Here's a quick look at each of these issues.

Quality of the Optics

Unless you can get your hands on some cameras and lenses, you can't really do much to compare optical quality from one camera to another to pick the best optics. We mention quality of the optics mostly because if we didn't, you might wonder what you need to know about this issue. You don't need to know much.

First, you may have heard that digital camera lenses don't have the same high optical quality as lenses in film cameras. That's essentially true, but not strictly true—at least, not if you have a high enough budget. You can buy a professional-level digital camera with the same camera body and lens system as professional film cameras. These have been available for years. They typically attach to a professional-level standard camera body as a *camera back*, literally becoming the back of the camera. The camera body takes all the standard lenses it would take for shooting film, which means that if you're willing to pay for it, you can have the same lenses as the best film cameras.

Lingo   A *camera back* attaches to a standard professional-level camera body to literally become the back of the camera. Put a digital camera back on a standard camera, and the result is a professional-level digital camera.

Consumer-level digital cameras—the kind we assume you're interested in if you're reading this book—are a different story. Early consumer digital camera lenses offered minimal optical quality, which is to say, the lenses were pretty awful compared to lenses in equivalent film cameras—no matter how you measured equivalence. The lenses in today's models are better, but still not a match for equivalent film cameras. That's not a problem as long as the lens in any given model is matched to the camera's ability to take advantage of it. You can usually trust manufacturers to get the balance right.

As the pixel resolution in cameras improves, you can expect lenses to improve along with it. The one thing you should keep in mind is that glass lenses are preferable to plastic lenses. They will generally give you sharper images, and they are less likely to get scratched or otherwise damaged. They also fare better in minimal lighting, doing a better job of maintaining brightness and focus around the edges of the photo. Other than that, for the most part, you don't have to worry about the optics.

Having said that, however, we also have to point out that there is one situation in which you should be concerned about the quality of the lenses. If you have, or plan to get, a camera with optional lenses that you expect to buy (for the reasons we discuss in the next section), the quality of optional lenses is much more of an issue than the quality of the built-in lens in the camera.

One of the reasons photographers have traditionally considered the lenses more important than the camera is because they can wind up costing far more. If you get enough lenses, it's easy to have a much bigger investment in lenses than in the camera. If the camera breaks, or, more likely with a digital camera, if you move on to a better one in two or three years, you'll want the option to get a later and greater model that can use the lenses you already have. That means you'll want lenses with good enough optical quality to let you get the best pictures you can with that later and greater camera.

Unfortunately, there's no easy way to look at a list of specifications and determine lens quality. The best way to find out about the quality is to look for current reviews of lenses designed for the particular camera you're interested in. There are a couple of rules of thumb to keep in mind, however, especially if you can't find any relevant reviews.

First, if you want to play it safe and have no other information to go on for picking a lens, consider staying with lenses from a company that's well known for its film cameras and lenses, like Nikon, Canon, or Olympus, and has a reputation to protect. Related to that rule, make sure you know who makes the optional lenses for a camera that you might be considering. If the lenses are offered as accessories coming from someone other than the camera manufacturer, check out the lens manufacturer's reputation for quality as well.

Changing the Lens

Lenses come in two varieties: permanent and interchangeable. As the names imply, permanent lenses are permanently mounted in the camera; interchangeable lenses are, well, interchangeable. You can take one off the camera and put another one on.

Even permanent lenses aren't necessarily unchangeable, however. Some built-in lenses are designed so you can add filters and converters (actually lenses) in front of the built-in lens. *Converters* are simply lenses that work with the built-in lens so that the combination gives you a different view, converting the camera's built-in lens into a wide angle lens, for example. This feature gives you the equivalent of interchangeable lenses, at least to the extent that you can change the way the image looks by the time it makes it through the built-in lens. We'll generally refer to converters as lenses, and, unless we say otherwise, when we talk about changing lenses, we're referring to both interchangeable lenses and converters.

> **Lingo** Some cameras let you attach options to the built-in lens to give it additional capabilities. Because these don't actually replace the lens, they are often called *converters*, as in *wide angle converter* or *telephoto converter*.

If you've never used a camera that let you change the lens, you might wonder why you would want to. The reason, in a word, is flexibility. As we mentioned in the section "What Kind of Photographs Do You Want?," different kinds of lenses let you take different kinds of pictures. If you fall in the mildly creative category, you might want the ability to switch from one kind of lens to another to match the lens to the picture, or series of pictures, you're taking at the moment. If you're a prosumer you almost certainly want that capability, almost by definition.

Even if you don't want to switch lenses a lot, you may need to change or add to a lens more or less permanently. If you're an architect who primarily wants to take pictures of buildings, for example, you probably would do best to get a wide angle lens. In any case, a camera that lets you change the lens can give you the flexibility you need.

Normal, Wide Angle, and Telephoto Lenses

Before we say any more about wide angle lenses and telephoto lenses, it will help to talk about something called a *normal lens*, which serves as a baseline of sorts to compare everything else to. A normal lens sees things the way you would see them, a definition that could use a little explanation.

Lingo   A *normal lens* lets the camera image match what the human eye would see.

The easiest way to understand what makes a normal lens normal is to look at mirrors. Lenses and mirrors actually have a lot in common. Both deal with light, and both can change the way a scene looks—something you've probably seen happen with mirrors during a trip to a funhouse.

A perfectly flat mirror shows a scene just as you would see it if you were looking at the scene directly. The *angle of view*, or *field of view*—what you see from left to right and up to down—is exactly the same as you would see if you were looking at the scene though a window the same size as the mirror. Everything is also in exactly the same proportions as you would see them if you were looking at the scene directly. In truth, except for the fact that you can see your own reflection, you really have no way to tell, just by looking, whether you're looking at a mirror or through a window.

Lingo   The *angle of view*, or *field of view*, of a lens is a measure of how much you can see, from left to right and up to down. Wide angle lenses have a wide field of view. Telephoto lenses have a narrow field of view.

What we've just described is a mirror with a normal view.

Mirrors don't have to be flat. Bend them, and you get funhouse mirrors. Depending on the curve, things can look fatter or thinner than they would if you looked at them directly, or proportions can be distorted. You don't even have to go to a funhouse to see the effect. One or more of the mirrors on your car probably has the message, *Objects in mirror are closer than they appear*, which really translates to *That car you're about to cut off is much closer than you think it is based on what you see in the mirror, so be careful.*

The reason cars and everything else look like they're further away in these mirrors is that the mirror isn't giving you a normal view. Instead, it's giving you a wide angle view, which means you are seeing more in the mirror than you would with a flat mirror. Being able to see more gives the impression of being further away, because with a normal mirror you would have to be further away to see it all.

A normal lens works just like a flat mirror, showing the same image your eye would see. A wide angle lens works like the car mirror, showing more than the human eye would see. A telephoto lens works like a telescope, giving you a much narrower field of view, but much more up close and personal.

Note   What counts as a normal lens depends on the camera and the size of the film or, for a digital camera, the size of the sensing element. A normal lens for a 35mm camera would be a telephoto lens for a digital camera, because the camera sensor is so much smaller than 35mm film.

As we mentioned earlier, the most extreme version of a wide angle lens is the fisheye lens, which will let you see a full 180 degrees in all directions. A somewhat less extreme wide angle lens will let you see, for example, the entire width of a building without having to step too far back from the building. A bit less wide angle is a good choice for landscapes or seascapes, or to give you the room for some extra people in a group photo. Telephoto lenses let you magnify the image from a distance. Figure 2-16 shows some variations of the same scene taken with lenses ranging from fisheye to telephoto.

Figure 2-16   Four views from the same spot, using four different lenses.

Zoom lenses, which we'll discuss in the section, "The Need for Zoom," will let you zoom the lens from one setting to another along the range from extreme wide angle to extreme telephoto. However, no lens will let you zoom over more than a small portion of the range. If you want to take different kinds of pictures, it still helps to have an assortment of lenses available. If you mostly take a particular kind of picture, you'll want to have the most appropriate lens for that kind of picture—which may or may not be the lens that comes with the camera, and may or may not be a zoom lens.

What Lenses Are Available?

Assuming you want the ability to change lenses, and the camera you have will let you change them, the next question is: what lenses do you want, and are they available for the camera? The easy answer is that you may want a wide angle lens, a telephoto lens, a macro lens (sometimes called a super close-up lens), a fisheye lens, or some combination of these, depending on what kinds of photos you want to take. But the real answer, of course, is more complicated.

We mentioned in an earlier note that what counts as a normal lens depends on the size of the sensor, which for a film camera means the film and for a digital camera usually means a *charged coupled device (CCD)*. (A few cameras use CMOS sensors, but CCDs are the norm, and anything we say about CCDs in this book will apply to CMOS sensors as well.) CCDs are much smaller than most film sizes, which means they have very different requirements for what counts as a normal, wide angle, or telephoto lens.

Lingo   Most digital cameras use a *charged coupled device (CCD)* in place of film to record images.

Because 35mm film is a long established standard, and CCD sizes are not, most manufacturers give ratings for digital camera lenses like *7mm equivalent to 38mm on a 35mm camera.* This is a useful piece of information if you're familiar with 35mm cameras. In case you're not, however, here's how to decode it.

The first thing you need to know is that camera lenses are rated in terms of something called *focal length*, which is measured in millimeters (mm). You don't need to know what focal length measures, except to know that it's related to the actual physical length of the lens, so a short lens has a short focal length, and a long lens has a long focal length. More important, a short lens gives you a wider angle of view and less magnification; a long lens gives you a narrower angle of view and more magnification. In other words, a short lens is a wide angle lens, and a long lens is a telephoto lens.

Lingo   The *focal length* of a lens tells you the angle of view and magnification. Short focal lengths give you wide angle lenses. Long focal lengths give you telephoto lenses.

The second thing you need to know is that for 35mm cameras, a normal lens is anything around 50mm. Anything much shorter than 50mm is a wide angle lens, and anything much longer is a telephoto lens.

Beyond these basics, everything is a matter of degree. A 200mm lens and a 400mm lens are both telephoto lenses, but the 400mm lens magnifies the image that much more. Table 2-3 gives some typical uses for various focal lengths (in terms of a 35mm camera).

Table 2-3   Appropriate Focal Length for Different Types of Photos

Type of Photo or Use	Appropriate Focal Length (with Typical 35mm Lenses)
Special purpose, fisheye photos: often used to get 180 degree view of tight spaces and interiors	15mm to 18mm
Wide angle: Landscape, seascape, room interiors, buildings with surrounding landscaping, large groups	24mm to 35mm
Normal: Midrange photos, snapshots, small groups	35mm to 55mm
Moderate telephoto: Close-ups from a moderate distance, allows candid shots without disturbing the subject	70mm to 100mm
Telephoto: Sports events, wildlife, distant subjects	100mm to 200mm
Extreme telephoto: Tight close-up in sporting events, surveillance, dangerous wildlife	Over 200mm

How Easy Is It to Change the Lens?

If you're going to get a camera that lets you change or add to the lenses, it helps to have one that lets you make the changes quickly and easily. Aside from the obvious convenience factor, getting a camera that can be a quick change artist will let you take shots that might otherwise get away. Some shots set themselves up, often thanks to specific lighting conditions. If you can't change the lens quickly, the light may change before you get the lens changed.

The standard for comparison is a bayonet mount, which lets you remove a lens by pressing one button and giving the lens a quick twist or mount a replacement by positioning it and giving another quick twist. Unfortunately, you won't find anything that quick and easy on consumer-level digital cameras. Most use lens attachments that have to be screwed on and off. Some, however, are easier to put on and take off than others. If you're planning to get lens attachments, it's worth trying them on the camera before buying to see how easy it is to mount the attachments. If you haven't bought the camera yet, you may have second thoughts if it's hard to change the lens. If you have bought the camera, you may want to reconsider investing even more money in lenses.

Adding Filters

The ability to add filters to a lens usually goes hand in hand with the ability to change a lens or add wide angle and telephoto converters. However, you don't need additional lenses or converters to take advantage of filters.

Filters come in all sorts of varieties. Some improve the camera's ability to see what you want it to. Polarizing filters, for example, reduce glare. Take a picture of someone swimming and you may see mostly reflected light in the photo. If you add a polarizing filter, the glare on the surface of the water will largely disappear.

Other filters let you create special effects. Star filters, for example, add a starburst effect to bright objects like the sun or streetlights. Still other filters, called neutral UV protective filters, basically give you a protective shield for an expensive lens. They protect the lens itself from scratches and other damage on the grounds that it's cheaper to replace a filter than a lens. If you consider your-self anything more than a point-and-shoot photographer, you'll want to consider using one or more filters, even if you have no intention of buying additional lenses. Even a point-and-shoot photographer may want to consider a polarizing filter.

The Need for Zoom

We've already touched on the benefits of zoom features, so we won't go over the same territory again. Besides, if you've ever used a zoom lens, whether on a still camera or a video camera, you already know its benefits: it lets you zoom in or out on whatever you're taking a picture of without having to physically move around.

Having said that, there are two important issues about zoom that we haven't covered yet.

First, don't assume that a zoom lens includes any part of the telephoto range. As we said when we were talking about changing lenses, a zoom lens will let you change the lens from one setting to another along some part of the range that goes from extreme wide angle to extreme telephoto. But there's no guarantee that any part of that range includes any magnification that qualifies as telephoto. To find out if the camera offers telephoto zoom, look for the two extremes of focal lengths. The ratings mean the same thing for zoom lenses as for lenses that don't zoom: a focal length that claims to be equivalent to 50mm on a 35mm camera is a normal view. Lower numbers indicate a wide angle zoom; higher numbers indicate telephoto zoom.

In fact, many, if not most, of the zoom lenses built into consumer-level digital cameras zoom from somewhat wide angle to mildly telephoto. The two photos in Figure 2-17 show a typical range for zoom. (These two happen to be taken with the Epson PhotoPC 3100Z.) The photo on the top shows the extreme wide angle view; the one on the bottom shows the same scene at the extreme telephoto setting. The range is equivalent to 34mm to 102mm on a 35mm camera.

Figure 2-17   Optical zooms for built-in lenses tend to hover around normal, going from mildly wide angle to mildly telephoto.

Even if you're a point-and-shoot photographer, you probably want a zoom feature. You can probably do without much in the way of telephoto zoom, although it would still be worth having. If you're a mildly creative photographer, you certainly want zoom, and almost certainly want telephoto zoom. If a given camera doesn't have telephoto capability as a built-in feature, make sure you can add it by changing the lens. If you're at the prosumer level, you probably don't even want to consider a camera without telephoto zoom, either built-in or available in an optional lens.

The other issue we haven't touched on yet is the difference between optical and digital zoom. Our advice on digital zoom is simple: ignore it. Pay no attention to whether a camera has the feature, and if you happen to notice it has digital zoom, make sure you never use it.

Let's beat this one into the ground a bit: Digital zoom is essentially worthless. It simply takes the image projected on the camera sensor and enlarges it.

You can get exactly the same result by taking the photo without digital zoom, enlarging it after you move it to your computer, and then cropping the image. There is no benefit to be gained by doing it in the camera instead. Ignore the feature. Case closed.

Macro Mode: How Close Do You Want to Go?

We've mentioned macro mode—the ability to focus to take a picture from just inches away instead of feet away—earlier in this chapter, including in the section "Extreme Close-ups." You should consider whether you want macro mode or whether you can live without it. Even if you're a point-and-shoot photographer, a macro mode can be useful. It will let you focus on a face, for example, from a closer distance than the typical 3 to 4 feet of cameras without a macro mode.

If you want a macro mode, consider how close you want to be able to get to take your pictures. As we mentioned earlier, some macro modes will limit you to taking pictures from several inches away. Others will let you focus from just an inch or two away. Keep in mind also that cameras that let you add attachments to the lens often will let you add a macro mode by way of an attachment. And some cameras with built-in macro modes let you add a macro lens for focusing even closer to your subject.   So if the camera you have doesn't let you get as close as you'd like to, check the available accessories to see if a close-up lens is an option.

What's SLR, and Why Does It Matter (But Maybe Not as Much as You Think)?

SLR is short for *single lens reflex*. In the world of 35mm cameras, it's pretty much the only choice for serious photographers, whether professional or amateur. The key feature that makes SLR cameras the cameras of choice is that when you look through the viewfinder, you're looking through the same lens that the camera uses to take pictures. That means you see exactly the same image that will get recorded by the camera.

Lingo   *SLR*, or *single lens reflex*, cameras use the same lens for viewing the image in the viewfinder and for taking the picture, so what you see in the viewfinder is exactly what you'll get in the photo.

The advantages of using the same lens for viewing the image and for taking the photo are easy to understand. If you've ever taken a picture that chopped off the top of someone's head, like the photo on the top of Figure 2-18, it was probably because the viewfinder showed you a different image than the camera saw. When you looked through the viewfinder, you saw the person's head, including the top, as in the bottom of the figure. You had no way to know that the camera would see the photo differently. Even if you knew it intellectually, you had no way to know how different it would look.

Figure 2-18 The scene the camera captures, as in the image on the top, doesn't always match what you see through the viewfinder, as in the image on the bottom.

The viewfinder and camera lens see slightly different scenes because of *parallax*, which makes an object appear to shift position when it's actually the observer, or in this case the lens you're looking through, that changes its position.

Lingo *Parallax* is the apparent shift in the position of an object when you look at it from a different position, like through a camera viewfinder compared to the camera lens.

If you're not familiar with parallax, you can see how it works easily enough. Hold a candle or some other object (your thumb will do) at arm's length, as in the photos in Figure 2-19. Close one eye and note the position of the candle compared to something in the background, like the tree near the center of the photo on the left of the figure. Now close the other eye (and open the first one) and you'll see that the position of the object appears to change in relation to the background, as with the photo on the right side of Figure 2-19.

Figure 2-19   Parallax gives each of your eyes a slightly different view.

With a camera, the distance between the viewfinder and the lens is enough to create a noticeable difference between what you see in the viewfinder and what the camera sees though the lens. Using an SLR camera eliminates this problem. Even better, in addition to showing you the exact scene that you're about to photograph, it shows you things like whether the image is in focus, because you're looking at exactly the image that the camera sensor will see.

The bad news is that in the digital world, SLR is confined to the higher end, more expensive cameras. The good news is that it's less important for digital cameras than for film cameras.

Most digital cameras include an LCD screen on the camera itself, and they let you use the LCD as your viewfinder to frame your shot. We hasten to mention that there are some drawbacks and limitations to using the LCD; it will drain your batteries more quickly, and you may not be able to see it in bright daylight. Despite that, however, the fact remains that with every camera we've ever tested, the LCD will show you the shot exactly as the camera will see it. So for photos that are close enough for parallax to be a potential issue, you can use the LCD to avoid chopping off the top of your friends' and family's heads.

We'd still argue that no matter what kind of photographer you are or what types of pictures you plan to take, you're better off getting a camera that shows you the same image in the viewfinder that the camera will see. But until prices for digital SLR cameras fall to a level that's within the bounds of reason, the camera's LCD will make an acceptable substitute.

Note   With SLRs, you often don't have the choice of framing the shot in an LCD. The image shows in the Viewfinder only.

Try This!   In the interest of complete accuracy we should mention that we've run across claims that the LCD preview and the final image don't match on some (always unspecified) cameras. We've never seen that happen—and we've tested dozens of cameras. Like they say, however, you can't prove a negative. So you might want to confirm the match between the LCD preview and the final image on your camera.

Testing this is easy enough. Carefully line up a shot with the edge of some object just inside the LCD view, then confirm it's just barely included in the captured photo. Since parallax may not be obvious at all distances, try this at several distances between, say, 3 and 10 feet. Also be sure to try it using both the top edge of the picture and the left or right side.

Choosing a Level of Control

The issue of how much control you need is straightforward. Just about any camera you can find offers automatic everything—automatic focus, automatic flash, automatic control of shutter speed, and so on. If you're a point-and-shoot photographer, the fully automatic modes are all you need and probably all you want. If you had to worry about options for manual settings, you'd probably find them more confusing than helpful.

On the other hand, mildly creative photographers will probably get frustrated—and prosumers will certainly get frustrated—if fully automatic mode is all the camera offers. The two functions of most interest are control over shutter speed and control over *aperture*, the opening that lets light through the lens and onto the sensor.

Lingo   The lens *aperture* is the opening that controls how much light gets through the lens.

Shutter speed controls how fast the camera can take a picture. If you can control the speed, you can do things like choose whether to stop action with a fast shutter speed, or give your photo a sense of blurred movement courtesy of a slower shutter speed.

Aperture controls *depth of field*, or how much of the image is in focus as measured by the distance to the camera. If you set your camera for a narrow depth of field, only part of the image will be in focus—as in the photo on the top in Figure 2-20, for example, with just the flowers in the foreground in focus. You can also set the depth of field to have more of the scene in focus, as in the photo on the bottom.

Lingo   *Depth of field* refers to how much of an image is in focus. With a narrow depth of field an object has to be within a narrow range of distances from the camera for it to be in focus.

Figure 2-20   Depth of field determines whether focus is limited to a narrow range of distances from the camera or includes a broader range.

You've seen this effect, and how it works to focus your attention on whatever's in focus, any number of times in films and TV shows. If you've never noticed it happening, keep it in mind the next time you go to the movies. If your camera lets you adjust depth of field, you can use the same trick for your photos, to help the viewer focus attention where you want it focused.

The reason for using a fully automatic mode for setting shutter speed and aperture is that the two interact with each other, and getting the right settings manually takes some thought. Both shutter speed and aperture affect the amount of light that gets used when you take a photo; you can get more light either by opening the aperture further or by using a slower shutter speed to keep the shutter open longer.

Unfortunately, you also have to find complementary settings for shutter speed and aperture, and those settings change with the lighting. For any given lighting, if the shutter speed is too fast for a given aperture, the image will be underexposed, meaning it will be too dark. Similarly, if the shutter speed is too slow for a given aperture, the image will be overexposed, meaning it will be too light. It works the other way too: for a given shutter speed, if the aperture is too small, it will give you an underexposed image. If the aperture is too large, it will give you an overexposed image.

Setting shutter speed and aperture manually can give you more control over how the picture will look, but figuring out the right settings is complicated enough that full manual control is best left to professionals and prosumers. Even those two groups will likely need full manual mode only rarely.

That said, mildly creative photographers and prosumers can make good use of both a shutter speed priority mode and an aperture priority mode. These are both automatic modes, but they give you control over one setting or the other. In shutter speed priority mode, you can set shutter speed, and let the camera figure out the right aperture to use. In aperture priority mode, you can set aperture, and let the camera figure out the right shutter speed to use. Just pick the mode that gives you control over the setting you care about most for the particular picture you're taking.

Other Features to Consider

The issues we've looked at so far in this chapter are the ones that we consider most important, but there are any number of other features that digital cameras offer. Here, in no particular order, is a quick look at other features of interest, along with short descriptions of each. We've mentioned some of these—like white balance and video mode—in discussing other features or types of photos. We've included those here to highlight them as specific features you should be aware of.

You can use this list either to decide which features you want, if you don't have a camera, or as a checklist of things to look for in the camera you already have. Don't worry for the moment about how to use any of these features. The point is simply to make sure you understand what features are available and why you might want them. We'll cover how and when to use most of these features over the next three chapters.

- **Metering modes**   In automatic mode, cameras set shutter speed and aperture based on built-in light meters. Depending on the photo and lighting conditions, you might want to base the settings on *spot metering*, which looks at a particular part of the scene—like the subject of the photo—or on a mode that looks at the entire scene. Having different metering modes lets you tell the camera which approach to use.

- **Sets of settings**   Some common categories of photos come out best with particular settings or clusters of settings. Some cameras let you set them all at once with preprogrammed sets of settings, with names like Close-up, Normal, Sports, and Landscape. Pick the Sports setting, for example, and the camera may use a fast shutter speed and a light metering mode that looks at the entire scene. Choose a Portrait setting, and it may use a wide aperture and spot metering. If your camera has this feature, it may or may not let you store your own custom set of settings. In any case, the more sets the camera has or allows, the better.

- **Confirm or delete each photo**   Most digital cameras include an LCD screen that you can use as a viewfinder. You can also use it to view all the pictures currently stored in your camera, making it easy to delete some to make room for more photos. A particularly useful feature is a confirm option, which automatically shows you each image after you take it and gives you a few seconds to decide whether to delete it or keep it.

- **Video mode**   One type of picture taking we discussed earlier was full-motion video clips. Cameras that offer this feature may let you take clips up to some number of seconds, or they may run a standard number of seconds for each clip. You typically won't have any control over the number of frames per second (too few frames leads to jerky video) or the resolution (which will always be less than you can use for photos).

- **Continuous mode**   Another type of picture taking we discussed earlier was a rapid-fire succession of stills. *Continuous mode* is one name for this feature. Another is *burst mode*.

- **Panorama mode**   If your camera has a panorama mode, it may simply change the aspect ratio of the picture, or it may provide help creating stitched panoramas. The section "What Kind of Photographs Do You Want?" covered both kinds of panoramas.

- **Sound recording**   If you want to take notes of your pictures to make it easier to keep track of them later, the ability to record sound will let you record the notes and keep them with each picture. Most sound recording features will let you record short messages—up to about 10 seconds or so. Keep in mind, however, that the storage space you use for sound notes is space you can't use for more photos. There ain't no such thing as a free lunch.

- **Changing exposure value**   Fully automatic camera modes base the various settings for the camera on an assumed exposure value. You may occasionally want to change this value for certain photos, so you can brighten or darken the subject of the photo without having to turn off the automatic mode. In short, cameras that let you change this setting give you a little more control over the photo.

- **Sensitivity setting or ISO setting**   For chemical photography, ISO (International Standards Organization) ratings indicate how sensitive a given film is to light. An ISO 400 film is more sensitive than an ISO 100 film. The higher the sensitivity of film, the less light you need to take a picture, but the more likely the picture will appear grainy. In digital cameras, the sensitivity setting, which is often given in terms of ISO equivalence, serves the same purpose. The higher the sensitivity, the less light you need, so you can, for example, take photos indoors without a flash. However, the higher the sensitivity, the more likely you'll see noise—like snow on a TV screen.

- **White balance**   We mentioned white balance when we talked about indoor photos. Briefly, colors change with changes in lighting, so things look different under sunlight compared to incandescent light. Adjusting the white balance to match the lighting ensures that white appears white and avoids a tinted look to the picture as a whole. Digital cameras in general adjust the white balance automatically. Some give you manual control, which you can use, for example, to get warmer colors under fluorescent light.

- **Television output**   Some cameras include a video output, so you can connect a video cable between the camera and your TV equipment (plugging into a VCR, for example). The feature effectively turns your photos into slides on a TV screen.

Key Points

■ Digital cameras don't fall naturally into a few neat categories. Each camera model has its own mix of features, so cameras that fall into the same category when you classify them one way fall into different categories when you classify them another way.

■ The best way to pick a camera is to start with the type of category that matters most to you, and narrow down your choices based on that type. Then look at the second most important type of category to winnow your choices down further, and so on.

■ The first step in understanding your needs is to know what kind of photographer you are (or want to be): point-and-shoot, mildly creative, or prosumer.

■ Also consider what kinds of photographs you want to take, and what features you need in the camera in order to take them.

■ If you never crop your photos and never print them at sizes larger than 5 × 7 inches, a 1-megapixel camera should serve your needs. If you print as large as 8 × 10 inches, you'll need a 2-megapixel camera instead.

■ If you sometimes crop your photos or print them at sizes larger than 8 × 10 inches, you want the highest resolution you can afford.

■ If you have more pixels than you need in a photo, you can reduce the number without hurting the image quality. If you have fewer than you need, you can increase the number, but the quality will not match a photo that was taken at the right pixel resolution to begin with.

■ Normal lenses see scenes much as the human eye sees them. Wide angle lenses see a wider angle of view. Telephoto lenses see a narrow angle of view, but the image is magnified.

■ A camera that lets you change lenses, or add adapters that serve the same purpose, will let you take a wider variety of pictures, from wide angle shots that will let you see more in the image without having to step back so far, to telephoto images that can give you a close-up view from far away.

■ Filters come in all sorts of varieties, including some that improve your photos (polarizing filters minimize glare, for example), some that add special effects to your photos (like adding a starburst effect to the sun), and some that simply protect an expensive lens from damage.

- Zoom lenses aren't necessarily telephoto lenses.

- Digital zoom is essentially worthless. Ignore it.

- You can use control over shutter speed to stop action, or choose to indicate movement with a blur courtesy of a slow speed.

- You can use aperture to control depth of field (how much of the image, as measured from distance to the camera, is in focus), which lets you focus attention on the subject of the photo.

- Shutter speed and aperture interact with each other, since both affect the amount of light that gets into the camera.

Chapter 3

Getting Started with Digital Photography

Getting started taking pictures with a digital camera can be as easy as putting in some batteries (you may have to charge them first), putting in the camera's storage card, uncovering the lens, and taking pictures. Virtually any digital camera you wind up with will have a fully automatic mode that will ensure the camera will take decent photos without you having to worry about them.

Relying on automatic mode, however, is not always your best choice. The vast majority of cameras will also have additional features you need to know about before you can take full advantage of the camera. At the very least, virtually any camera should give you a choice of settings that will help determine the resolution of the picture—both in the sense of how many pixels it will have and how well it can resolve fine detail. Keep in mind also that after you take the picture, you'll have to figure out how to do something with it—move it to your computer, e-mail it, print it, or simply delete it.

What follows is a whirlwind tour of all of these things. It's meant as an introduction to what you need to know to get started with point-and-shoot digital photography. We'll cover the most common settings that most cameras offer and talk a little about how—and in some cases whether—to use them. Your camera may not offer all the features we'll cover here, but for those it does offer, you'll find useful information about how to take advantage of them. The mechanics of using the features will vary from one camera to another, so we won't provide details about how to change settings. If it's not obvious from looking at the camera itself, you should be able to find the information in your camera's manual.

We assume at this point that you've already worked through the obvious steps in the camera manual for getting started—like charging the batteries if they're rechargeable, loading them, and inserting the camera's storage cards, often called *digital film*. We'll also assume that you've gotten as far as learning the mechanics of working though the camera menus and learning which are the right buttons to push. We'll pick up from there.

Lingo   Camera storage is often referred to as *digital film* because it's analogous to film in a film camera.

Common Features and How to Use Them

Most cameras come out of the box with everything set to use automatic modes—automatic flash, automatic white balance, automatic exposure, and more. The idea is to make sure that anyone can pick up the camera and get pretty good pictures without having to think about it. If you're willing to invest just a little time exploring the features, however, you'll take much better pictures. There are some features you need to set before you can use them at all, and there are some that may be on by default that you should turn off and ignore. Here's a strategy for getting familiar with your camera, covering the important points in the order you should learn them.

First Things First

You may have dived right in to take pictures already, and even gotten some good results. But if you can't answer questions like, "What's the minimum distance you have to be from your subject?" you should get the answers before you take your next picture. Mistakes aren't as expensive with digital cameras as with film cameras—at least you don't have to pay for film and for developing shots that don't come out. But there's no point wasting time taking pictures that are bound for deletion. So let's start with the basics.

Caution   High on the list of first things: don't remove the memory card from your camera, or put it back in, without turning off the camera first. Otherwise you risk scrambling the files currently stored on the card.

Viewfinder and LCD

This sounds almost too obvious to mention, but check out the viewfinder and liquid crystal display (LCD). The viewfinder first. It's amazing how many people—including accomplished photographers—don't realize that cameras often give you a way to focus the viewfinder. Or maybe this isn't too surprising, considering that the information may be missing, or very well hidden, in the camera manual. If your camera lets you adjust the viewfinder, you can correct for nearsightedness or farsightedness, which will let you use the viewfinder without glasses and still see a clear image.

Note   Where's the viewfinder? Not all digital cameras have viewfinders. The Nikon Coolpix 2500, for example, relies entirely on its LCD for framing the picture.

In any case, the feature is properly called a *diopter correction*, although you'll rarely see it called that. We'll just call it a viewfinder focus control. If you wear glasses, or if the view through the viewfinder looks blurry, look to see if there's a small knob or wheel somewhere in the immediate vicinity of the viewfinder. Figure 3-1 shows the viewfinder focus control on an Epson PhotoPC 3100Z, for example.

Lingo   A *diopter correction* on a camera lets you adjust the view through the viewfinder so you can use the viewfinder without glasses.

If you find something that looks like it may be a viewfinder focus control, you may be able to find a reference to it in the manual to confirm what it is. But it's probably faster to look though the viewfinder and turn the control to see if it affects the image. If it is a viewfinder focus control, set it to give you a sharp image, and you'll get a better view of the pictures you're taking. (And now that you're aware of the possibility, if you pick up a camera and notice the image though the viewfinder is blurry, you know how to fix the problem.)

Figure 3-1   Look for a small focus control near the viewfinder.

While you're at it, check the LCD also, assuming your camera has one. Make sure you know where the control is that turns the LCD on and off. You'll often find it on a dial that sets the camera mode and is billed as a choice between using the LCD (LCD mode) and the viewfinder (viewfinder mode). On every camera we've ever seen, however, all the viewfinder mode does is turn the LCD off, just like the dedicated LCD on–off button that you'll find on other cameras. As you might guess, you can use the viewfinder even when the camera is set to use the LCD. However, you should get in the habit of turning off the LCD when you're not using it; your batteries last longer that way.

Also find out if there's a way to adjust the LCD brightness. You may find it helpful to turn up the brightness so you can see the LCD more easily in bright light, or turn it down to stretch out battery life when the light is low and you can see it easily.

Zoom

If the camera has a zoom control, you'll want to check that out too. Look in the manual, if necessary, to find out how to zoom—typically you press one button to zoom in and another to zoom out. More important, check the manual to see if there is a digital zoom, and, if so, if there's a way to turn it off. If there is, turn it off and forget that the feature is there. As we mentioned in Chapter 2, "Knowing (and Choosing) Your Camera," digital zoom lowers the resolution of your photo, and it doesn't do anything for you that you can't do by cropping the picture later. There's no benefit to using it, and if you don't turn it off, you may use it accidentally and wind up with a lower resolution than you want.

If there's no way to turn off the digital zoom, at least find out how it works, and how you can tell when the zooming switches from optical to digital. (This assumes, of course that the camera has an optical zoom. Some cameras offer digital zoom only.) The camera may zoom to the extreme of its optical zoom, for example, then wait a full two seconds before continuing on with digital zoom. Depending on how obvious the switch is, you may find that it's easy to avoid using the feature. If the camera has digital zoom only, our advice is not to zoom at all. Leave the camera in its unzoomed state and ignore the feature.

Assuming your camera includes an optical zoom, if you don't already know the camera's focal length range from one extreme of zoom to the other, you should look it up. The focal lengths will tell you if the camera has a wide angle or telephoto capability, and if so, how much of a wide angle or telephoto feature it offers. You can see the range for yourself by zooming the lens and looking at the image on the LCD, but it's useful to know the numbers, if only to understand how the range you have available fits into the wider context of wide angle and telephoto possibilities.

You may see the focal length printed on the lens casing itself, as in Figure 3-2. As we mentioned in Chapter 2, however, the actual focal length will mean something different in practical terms depending on the size of the sensor in the camera.

Figure 3-2   What you need to know is not the focal length printed on the lens as shown here, but the 35mm equivalent.

A better place to look is on a specifications page in the manual, where you should find both the focal length of the lens and what the 35mm equivalent is. Remember, for 35mm cameras, 50mm is a normal lens that sees the world pretty much as you see it. Anything much below that is a wide angle lens. Anything much above that is a telephoto lens. For more details, be sure to compare the focal lengths to the ranges we talked about in Chapter 2, in the section "What Lenses Are Available?"

How Close Can You Go?

It's critical to know the minimum distance you have to be from a subject to take a picture, which is to say, how close you can get to something and still be in focus. If your camera has a macro mode, there will actually be two closest distances, one with macro mode off and one with it on.

Look for a section in the manual telling you how to take a picture. It should include a warning about how close you can get without turning on macro mode (assuming the camera has one), and how much closer you can get with macro mode. If you can't find the information anywhere else, turn to the camera specifications and look for focus range, shooting range, or some similar name. By whatever name, the range will run from some number to infinity, with the first number varying widely. The numbers for four cameras whose manuals we have handy as we write this, for example, are 11.8 inches, 20 inches, 20 inches (again), and 2 feet. Still other cameras have to be as much as 3 or 4 feet from the subject.

Macro mode, which some cameras call close-up mode, will let you focus at much closer distances. Here again, you'll see wide variations. The closest distances for the same four cameras we just referred to are 1.6 inches, 2.3 inches, 3.9 inches, and 9.6 inches.

Whatever the numbers for your camera, find out what they are and keep them in mind when you're taking a close-up. If your camera has a macro mode, make sure you know the distance at which you have to switch from standard mode to macro mode. You might even want to write these numbers on a label and stick it on your camera or camera case.

Find Out How to Frame the Shot

If you use the camera's LCD to frame your pictures, framing is not a problem because you're looking at the scene as the camera sees it. (As we pointed out in Chapter 2, we've seen claims that this isn't always true. We can't prove a negative, but we haven't seen a problem in any of the dozens of cameras that we've looked at.) Similarly, if you have one of the few digital cameras available that uses the same lens for the viewfinder that the camera uses for taking the picture, framing is not a problem. You look through the viewfinder, and whatever you see, once again, is what the camera will see.

Unfortunately, you may not be able to use the LCD in bright daylight, because it's too washed out. Or you may prefer turning off the LCD to extend battery life. Either way—assuming you have one of the vast majority of cameras, with a viewfinder that doesn't look though the camera lens—you have a problem. Look through the viewfinder, and you'll see a slightly different image than what the camera sees. (We discussed this issue in some detail in Chapter 2, in the section "What's SLR, and Why Does It Matter (But Maybe Not as Much as You Think)?"

The exact difference between the views will depend on where the viewfinder lens is in relation to the camera lens and on how far you are from the subject when you take your picture. It's helpful to have some idea of the difference so you can compensate for it and avoid chopping off the top of someone's head.

Check your camera's manual to see if there's any useful information about the difference between the views. If you're lucky, you'll find something that tells you the distance to the subject that will give you the same view in both. For example, the two may be set to match at, say, 10 feet. If you're closer than that, the top of the scene as you see it in the viewfinder would be chopped off in the photo. If you can find that information you'll know when to leave extra room at the top of the frame. (Of course, then you'll have to remember to do it, but that's another issue.)

If you can't find this information in the manual, you can figure it out quickly for yourself, thanks to the instant availability of digital photos. It won't even cost you anything, since you can delete the test shots when you're done. (Not so incidentally, film cameras that aren't SLR have this same problem, but to run this test on a film camera, you'd have to pay for developing the film. More important, non-SLR film cameras don't have an LCD to use when framing really matters.)

First, be aware that the difference between the two views will only show up in the direction or directions that the viewfinder lens is offset from the camera lens. If the viewfinder lens is immediately above the camera lens, so the centers of both are lined up parallel to the side of the camera, the only offset in the view will be in the up and down direction. If the viewfinder lens is both above the camera lens and offset to the side, the offset in the images will be both left to right and top to bottom. For this discussion, we'll assume your camera lens and viewfinder are lined up top to bottom. If they aren't, you'll have to take the left to right offset into account also.

Find something you can take a picture of that includes an obvious horizontal line, like the frame of a painting or a door frame. Then measure distances at

2-foot intervals from 2 feet away from the subject (assuming the camera lets you focus from 2 feet away) to 14 feet away. (You can use a tape measure if you want to be exact, or you can be more casual and just count paces.) Starting at the 2-foot distance, take a picture with the top edge of the horizontal line just barely inside the viewfinder frame. Then step 2 feet further away and take the same picture again. When you're done, you should have one picture that's a close match for the view in the viewfinder, with the top of the frame just barely in view at the top of the photo. That tells you the distance that gives you a matching picture. The other shots will give you a sense of how much extra room you need to allow at the top of the viewfinder at different distances.

Keep in mind too that if the offset is, say, top to bottom when you're holding the camera horizontally (so the long side of the picture is parallel to the ground), it will be offset to the side when you're holding the camera vertically. Figure 3-3, for example, shows two shots taken at 2 feet and 10 feet from a lamppost. In both cases, the lamppost was centered in the viewfinder. In the version taken at 2 feet, it is noticeably off to one side.

Figure 3-3   When we took these pictures, the lamppost in both cases was centered as seen through the viewfinder.

Having said all this, note that if you're taking a picture that you really care about framing just so, you should use the LCD to ensure that you're doing it right. And if you're using macro mode, you should absolutely, without question, use the LCD to frame your picture.

Half Measures: Auto-Focus and Auto-Exposure

If you're using the auto-focus and auto-exposure settings—and you're well advised to, at least while you're learning your way around the camera—make sure you know how to use them. With virtually any camera today, you aim the camera at the subject of the picture, and gently push the shutter button part way down. The camera then focuses the lens, determines the exposure, and locks in the settings.

The camera will give you some indication when it's ready to go on. For example, if you're looking through the viewfinder, you may see a blinking light turn to a solid light. If you're looking at the LCD, you may see an icon change shape or appear out of nowhere. (You'll have to check the camera manual to find out what you're looking for.) Once the camera indicates that it's ready, you can press the button the rest of the way to take the picture.

There are a couple of tricks that will let you take best advantage of these automatic features. First, keep in mind that the automatic focus and automatic exposure features assume that the item you want in focus and properly exposed is in the center of the picture. This may or may not be true.

Suppose, for example, that you want to take a picture of a wooden fence post. You can frame the picture with the post in the center of the frame, as in Figure 3-4. That's what the auto-focus feature assumes you're doing, and it will focus on the post precisely because it's in the center of the frame.

Figure 3-4   The fence post is the intended subject of this photo, but putting it dead center makes for a boring shot.

Sometimes, however, you can get a more interesting photo if you move the subject of the photo off center. Unfortunately, if you frame the picture this way, then press the shutter button halfway down, the camera will automatically focus on whatever's in the center of the field of view, which can leave the actual subject of the photo out of focus, as in Figure 3-5.

Figure 3-5   If you reframe the scene so the fence post is off center, auto-focus will focus on the wrong thing.

The workaround for this is simple. Frame the picture as in Figure 3-4, press the button halfway down until the camera locks in its settings, then reframe the picture as in Figure 3-5, and take the picture. Figure 3-6 shows the result, with the picture focused as in the first version, but framed as in the second version.

Figure 3-6 By letting the camera focus first and then framing the scene, you can get both the framing and the focus you want.

The same basic trick applies to getting the automatic exposure set properly for the subject of the picture. Suppose, for example, that you want to take a picture of a lamp that happens to be near a window, and you want to frame the picture in a way that happens to put the window near the center of the screen. The top of Figure 3-7 shows what happens if you frame the image first and then hold the button halfway down. The scene outside the window is nicely exposed, but the lamp that's the subject of the picture is much too dark. In the photo on the bottom of the figure, we first set the exposure by pointing at the far right side of the lamp—away from the window—and holding the shutter button halfway down. Then we reframed the picture and took the shot.

Figure 3-7   You can change the way a photo looks by changing where you point the camera while the auto-exposure feature determines settings.

A Step Beyond Basics

As you learn to use the camera, there are a few other things you'll want to take a look at sooner rather than later. High on the list are the camera's flash settings and preprogrammed clusters of settings for specific kinds of photographs. We'll tackle the preprogrammed sets of settings first.

Sets of Settings

Not all cameras have this feature, which goes by various names, including scenes, programmed settings, and best shot mode. For cameras that offer the feature, you may find only a few choices, or a long list. Whatever the details, if your camera includes this feature, you'll want to know what the settings are and how you set them. If the manufacturer chose the settings well, the easiest way to take advantage of the camera's features will be to understand which setting is best for which kind of photo. Using these preprogrammed settings will also be the fastest way to change to the right settings when you're about to take a photograph.

Different cameras use different names for similar purposes—not just for the feature overall, but for each set of settings. That said, typical choices for settings include the following:

- Sports (for action shots)
- Beach or Snow (for bright scenes with lots of reflections)
- Sunset (to bring out the warm colors in sunsets and sunrises)
- Landscape (for landscapes and seascapes)
- Night Landscape (for pictures of a fully lit city skyline, for example, or well-lit scenes like an amusement park at night)
- Museum (for indoor shots where flash is not permitted, or where you don't want to use flash for some other reason)
- Fireworks (for fireworks displays)
- Close-up
- Backlight (for pictures with light coming from behind the subject)
- Copy (for taking pictures of black lines on a white background, like text on a page or notes on a whiteboard)
- Portrait (for taking portrait shots of people)

Also check to see if you can define your own clusters of settings. You probably won't want to define any right away, but you may want to as you get more familiar with the camera. You might as well find out now whether you can or not.

Try This! In some cases, the camera's manual will tell you only what kind of scene each mode is meant to shoot. In other cases, however, it will also tell you the settings for each mode. If you can find out the individual settings the camera uses for each type of photo, look up the purpose of each setting as explained in the camera's manual or in this book (or both) and see if you can figure out why the camera manufacturer chose those settings.

Even if the manual doesn't tell you what settings a particular mode uses, you may be able to figure out at least some of the settings. For example, with one of the cameras we used while writing this book, we know that a particular choice set the flash to a particular mode, because the flash icon changed to that mode every time we went to that choice. You can also try the manufacturer's Web site for more details. If they aren't posted, look for a link for technical support, and ask what settings change with each mode.

The point of this exercise is not just to solve a puzzle. If you understand why the preprogrammed settings are set the way they are, you'll go a long way to understanding when to change the settings yourself for other kinds of pictures, when and as needed.

Flash

Most cameras today have a built-in flash, and virtually all cameras with this feature offer several modes. You want to know what modes are available, both because there are some you might want to use right away, even for outdoor shots, and because you'll want to keep some of them in mind for special situations even if you have no need for them right away.

The most common modes are auto, forced flash (also called fill-in flash), red-eye reduction, and off. Many cameras offer additional settings as well. Among the cameras we gathered for writing this book, for example, the Epson PhotoPC 3100Z and Olympus D-380 offer variations on something called slow synchronized flash (although the Olympus camera uses a different name for it). Here's what you need to know about each of these settings.

Auto is usually the default setting, and it will be the right choice for many, if not most, photos. With auto-flash, the camera uses the flash when it decides it's needed, and it doesn't use the flash otherwise. If you leave the camera at this setting, you'll usually get good pictures, but you can get better pictures if you're a little smarter about what you do and use the other modes when they're called for.

Forced flash will tell the camera to flash whether there's enough light or not. This feature is often called *fill-in flash* because if you use flash in some situations, like taking a picture outside in bright sunlight, the flash will fill in shadows. The photo on the top of Figure 3-8, for example, was taken without flash in sunlight, with the sun in a position where it's casting harsh shadows. The photo on the bottom is the same scene with the camera set to flash, to fill in light and lessen the shadows.

Lingo   *Forced flash*, also called *fill-in flash*, tells the camera to use the flash whether there is enough light available or not. The extra light can fill in shadows that would otherwise ruin the picture.

Figure 3-8 Fill-in flash can help minimize shadows falling on your subject.

Forced flash can also come in handy if you are taking a picture of someone, or something, with the light behind the subject. Probably the first thing most people learn about outdoor photography is that you're not supposed to take a picture of someone with the sun behind him or her. The left side of Figure 3-9 shows what happens if you do.

The situation is known as *backlighting*, because the light is in back of the subject. What often happens with a backlit subject is that the camera will adjust its exposure setting for the bright light behind the object you're taking a picture of. The object itself will wind up dark, if not completely black, as in the photo on the left side of Figure 3-9. With forced flash, the flash will provide enough light so you can see the front of the object, as in the right side of the figure.

Lingo   With *backlighting*, the light source is behind, or in back of, the subject of the photo.

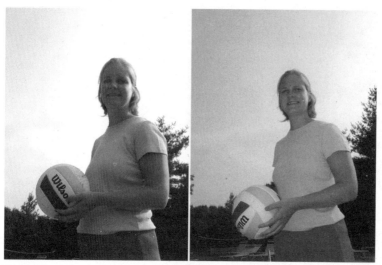

Figure 3-9   Take a picture with the sun behind the subject, and your subject may appear dark, as in the picture on the left. You can use flash to light the subject, as on the right.

Backlighting can be a problem indoors too, as in the photo on the top of Figure 3-10. Here again, you can use forced flash to light the scene, as in the version on the bottom of the figure. This scene is similar to the one we used to discuss the auto-exposure setting, with light coming through a window that's in the middle of the scene you're taking a picture of. You may need to use both tricks at the same time to get the best picture—forced flash plus pointing the camera away from the window while it's taking its measurements. However, they are different approaches to solving similar problems.

Figure 3-10   Use forced flash to light up the front of a subject that is lit from behind.

Red-eye reduction is one of those features that sounds great but causes more problems than it solves. We strongly recommend staying away from this setting.

Briefly, eyes do a great job of reflecting light. If your cat ever gets out at night and won't come when you call (a normal state of affairs for a cat), try walking around and shining a flashlight into the bushes. If your cat is in the path of the light and is looking your way, the light will get reflected back to you, and the glowing eyes will stand out like two gigantic fireflies next to each other. Human eyes do essentially the same thing.

This reflection can be a problem when you're using a flash. The iris, which opens and closes to let more or less light into the eye, is going to be nearly wide open if it's dark enough so you need a flash. Press the shutter button and you have lots of light going into the eyes and getting reflected back to you. Fairly often, this results in the camera seeing the eyes as red, because that's the color being reflected back. That's the *red-eye effect*.

Lingo   The *red-eye effect* comes from light—usually a flash—being reflected from a subject's eyes back to the camera.

One way to get rid of red eye is to force the iris in your subject's eye to shut down, and make the pupil, which is the opening into the eye, smaller before taking the picture. With the iris closed down, there won't be as much light getting into the eye or reflecting back to the camera. The red-eye reduction setting for a flash accomplishes that by setting the flash off before taking the picture, giving your subject's eye a chance to react, and then, finally, taking the picture.

Unfortunately, there's a problem with trying to fix red eye this way. By the time the camera finally gets around to taking the picture, it may not be there any more. The subject may have moved or simply changed his or her expression from the one that you wanted to get. Worse, if you accidentally leave the feature on for pictures that don't need it, the possibility of losing a shot is even greater.

Say you've been taking pictures of people at a party with the red-eye feature on, and now want to get a shot of the birthday girl blowing out the candles on the cake. If you forget to turn off the red-eye feature, there will be a significant lag between pressing the shutter button and taking the picture. The candles may all be out by then.

The first thing you should know about red eye is that you can avoid it without the red-eye reduction feature. Tell your subject not to look directly at the camera, turn up the lights so the subject's irises will close down, or both.

The other thing you need to know is that one of the real advantages of using a digital camera is that you don't have to worry about red-eye very much. That's not because a digital camera is any better than a film camera for avoiding red eye—it isn't. However, fixing red eye after you move the photo to your computer is trivial. Most photo editing programs even have a simple command to remove red eye, as shown in Figure 3-11. Select the area around the eyes, choose Remove Red Eye, and the problem is gone.

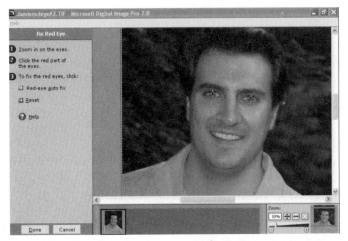

Figure 3-11 Removing red eye is as easy as selecting the area around the eye, as shown here, and giving the command to remove red eye.

Not so incidentally, not all cameras even bother with a red-eye reduction mode. And by our lights (there's a pun in there somewhere), you're better off without it.

Slow synchronized flash is a more sophisticated flash mode that comes in two basic variations. Both versions keep the shutter open for longer than the flash lasts (that's the slow part), but in one version the flash comes at the beginning of the process of taking the picture. In the other version, it comes at the end. More important, the two approaches yield different results visually and are meant for very different purposes.

Lingo   *Slow synchronized flash* is a less common flash mode that keeps the shutter open for a relatively long time, either before or after setting off the flash.

One version of slow synchronized flash starts out with a flash, then keeps the shutter open for a while longer. It's meant to solve a common problem for taking pictures outdoors at night. As the photo on the top of Figure 3-12 shows, if you use a normal flash mode for this sort of picture the background will be relatively dark. In many cases, in fact, your subject will be surrounded by black. The photo on the bottom shows the same scene using slow synchronized flash with a leading flash. The leading flash first captures the subject, and then the background fills in slowly over the next few seconds or few tenths of a second. As you can see, the lit area inside the house in the background is lit much more brightly in the photo on the bottom.

Figure 3-12   Slow synchronized flash with leading flash lets you capture both your subject and the background at night.

This slow synchronized mode with a leading flash is so tied to taking pictures at night that Olympus identifies it as Night Scene flash mode on its cameras. If your camera includes this flash mode by whatever name, you should mount the camera on a tripod when you use it. It's important that the camera stays steady while it's filling in the background. Otherwise, you may get a blurry picture.

Slow synchronized flash that ends with a trailing flash is a different animal altogether. It's basically a way to get a special effect as part of the picture, with the subject in your photo surrounded by one or more blurred objects. Figure 3-13 gives you some idea of what you can do with the version with trailing flash.

Figure 3-13   Slow synchronized flash with trailing flash lets you create some special effects in the camera, rather than adding them later by editing the photo.

As you can see in the figure, slow synchronized mode with a trailing flash lets you capture a blur of movement, which in this case is a train rushing past a platform in a train station, and then end with a flash to freeze the subject of the picture. In this photo, you can see both the streak left by windows as the train went by the camera and a frozen shot (from the flash) of the train emergency exit, a door, and some ghostly individual windows. The resulting photo gives a sense of movement overlaid with a frozen moment in time.

Note that you don't necessarily need to use a tripod with the trailing version of slow synchronized flash, because the blur from camera movement can actually enhance the effect. In the photo in Figure 3-13, for example, you can also see some blurred lights, vaguely resembling question marks, caused by camera movement while the shutter was open.

Ultimately, any version of slow synchronized flash is a relatively sophisticated option that you probably won't have reason to use right away. But if your camera offers one or more variations on the mode, tuck that piece of information into the back of your mind. You might want to experiment with the feature once you are otherwise familiar with the camera.

Macro Mode

Using a macro mode, if your camera offers it, is reasonably straightforward: when you're too close for normal mode, switch to macro mode and take the picture. There are a couple of special considerations about macro mode that aren't immediately obvious, however.

We mentioned earlier, when discussing framing, that for cameras whose viewfinders do not look through the camera lens, you should always use the LCD to frame your pictures in macro mode. That's because the difference between the view in the viewfinder and the view through the lens grows greater as you come closer to the object you're taking a picture of. The more extreme the difference, the harder it will be to know what the picture will look like by looking through the viewfinder. Using the LCD eliminates this problem.

The second issue is lighting. Some camera manuals warn that you shouldn't use the built-in flash with macro mode because it may wash out the image. In truth, if you take a picture from just a few inches away, that's the least of your problems. The camera itself may be blocking the light, so the subject of the picture is in the camera's shadow. And depending on the shape of the camera, the lens may sit between the flash and the subject, which would block the light from the flash as well.

Some cameras offer special lighting options for macro photography. Figure 3-14, for example, shows an optional accessory that works with most of Nikon's Coolpix cameras. Basically, it's a ring of lights that fits around the lens to light the area in front of the lens.

Figure 3-14   Nikon's Macro Cool-Light SL-1 ring light offers an easy way to light close-in macro shots. (Photo courtesy of Nikon, Inc.)

An option like this may not be available for your camera, however. And there's little point in investing in something like this if you won't need it very often. For taking an occasional close macro shot, you can improvise. For example, we took many of the macro shots for this book outside to take advantage of daylight lighting. The only trick was to be careful not to let the camera shadow fall on the subject of the picture.

One last consideration for a macro shot is that the closer you get to your subject, the more important it is for the camera to be rock-solid steady to make sure the details are sharp and clear. Most often, that means using a tripod. Also, if the camera has a zoom, be sure to check the camera manual to see if you have to set it to any particular position, like zoom it to the extreme wide angle position.

Getting into the Deep End

Finally for features, once you're fully comfortable with everything we've discussed up to this point, take the time to read the manual and find out what else your camera offers. We mentioned many of the possibilities—like continuous mode, video mode, and panorama mode—and discussed why you might want to use them in Chapter 2, in the section "Other Features to Consider." We won't repeat that information here, but you should go back to that section and take another look at it when you're ready to delve into more advanced features. We particularly suggest looking at our comments on metering modes, white balance, exposure value, and sensitivity settings.

There are several other advanced features we haven't touched on yet. One deserves special mention and a section of its own: sharpening.

Sharpening in the Camera

In digital photography, sharpening is an effect that detects edges—as the boundary between dark and light areas—and enhances them so they are easier to see. Figure 3-15, for example, shows an image with different levels of sharpening, starting with no sharpening in the photo at the upper left of the figure, the levels are 150 percent (upper right), 300 percent (lower left), and 500 percent (lower right). As you can see in the figure, the apparent gain in sharp detail can be significant.

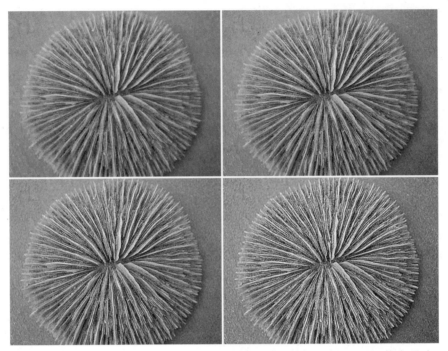

Figure 3-15   Digital sharpening in the camera compensates for blurriness in compressed photos by enhancing edges at the boundaries between dark and light areas.

Digital cameras generally sharpen the images in the camera as a matter of course for any images that you take in compressed format (we'll cover compression a little later in this chapter). The feature maintains, or restores, the sharpness of edges that otherwise get a little blurred because of compression.

Unfortunately, sharpening also tends to increase the visibility of jaggies and other flaws, or *artifacts*, that compression adds to a photo. In recognition of that issue, some cameras give you control over how much sharpening they do. The Nikon Coolpix 2500, for example, will let you set sharpening to automatic, high, low, normal, or off. As you get more familiar with your camera, and more sophisticated about using it, you might want to experiment with different sharpening levels if you have the choice, or turn off sharpening altogether and do your sharpening in your graphics editor, where you can undo it if you don't like the result.

Lingo   *Artifacts* are visible flaws added to an image by digital processing, like jagged edges that weren't jagged in the original scene.

Other Advanced Features

Here is a list of still more features you may want to explore. Some of these—like a timer—are available in most cameras. Others—like shutter speed priority and aperture priority modes—are relatively rare, but well worth having if you want to do more than point and shoot. All are worth taking a look at once you are thoroughly familiar with your camera's basics:

- **Shutter speed priority mode and aperture priority mode**   If you care about controlling whether the camera freezes a shot or creates a blur (which may be the more desirable choice if you want to show motion, for example), you may want to use a shutter speed priority mode for a given picture and let the camera automatically figure out the aperture to use. If you care about controlling the depth of field (because you want to make sure everything is in focus for a given shot or ensure that the foreground or background is not in focus) you'll want to use an aperture priority mode, if available, and let the camera figure out the shutter speed setting to use.

- **Manual focus**   Ever try to take a picture of, say, a horse in a field, wait for the perfect shot, frame it, and then have the horse gallop away before your camera focuses? If you set the camera for a particular focal distance, such as 3 feet, 10 feet, or infinity, you won't have to wait for the camera every time you take a shot. This will let you take pictures more quickly, with less chance of losing the shot because it's not there anymore.

- **Timer**   Want to take a picture of yourself? Timers let you take pictures a given length of time after you press the shutter button, waiting 10 seconds, for example. This lets you press the button and then get in front of the camera to include yourself in the picture.

- **Interval shooting with a timer**   Some cameras let you set them to take a picture every so many seconds, minutes, or hours. The result is a series of time-lapse stills.

Choosing Resolution and Compression Settings

Someday, memory will be cheap enough and camera processing fast enough to let you take all your pictures at resolutions we can only wish for today. When that day comes, you won't have to worry about how many shots can fit on your digital film. Until then, however, you have to make compromises. And that means making decisions.

To have any chance of making smart decisions, you need (among other things) to understand the difference between a camera's resolution setting—the number of pixels it uses—and its *compression* setting, or how much it will compress the image to save space on the storage card. We've already covered resolution a bit in Chapter 2, in the section "Choose a Resolution: How Much Do You Need?" Let's start here with compression.

Lingo　*Compression* lets cameras store photos in smaller files than they would otherwise use, so you can fit more photos on the storage card.

Compression: Lossless and Lossy

The first thing you need to know about compression is that there are two fundamentally different kinds of compression: *lossless* and *lossy*. The difference between the two is just what the names imply: With lossless compression, you don't lose any information. With lossy compression, you do.

Lingo　*Lossless compression* retains all the information from the original file. *Lossy compression* loses some information.

More precisely, when you decompress information that's been compressed with a lossless compression scheme, you get every single bit back. And we mean *every bit* in both the generic sense and the technical sense. If you do a bit-by-bit comparison, the before and after versions of the file will match exactly. On the other hand, when you decompress information that's been compressed with lossy compression, you lose some information. If you do a bit-by-bit comparison with the original file, the before and after versions will not match.

Whether you realize it or not, you may already have some experience with both kinds of compression. If you've ever used a zip file format to compress a file before sending it by e-mail, or if you've unzipped a file, you've used lossless compression. If you've taken any pictures with your camera yet, you've almost certainly used lossy compression.

The two kinds of compression serve different purposes. Some kinds of files—programs, spreadsheets, word processing documents, and the like—can't tolerate losing any information. Drop a few bits, and your program won't work, your critical spreadsheet for tomorrow's meeting will be scrambled, and your great American novel will no longer open in your word processing program.

Unfortunately, the requirement that you be able to fully reconstruct the original file limits how much you can compress the file. There's no point in getting into how compression works. Suffice it to say that there are all sorts of tricks that will let you drop file size. Even so, you will eventually reach a point where any further compression will lose information. On average, most compression schemes can cut files to roughly half their original size before you get to that point. That's nowhere near enough to do you much good for storing photos.

However, photos are a different kind of data. If you lose information from a photo file because of compression, you won't be able to exactly reconstruct the original file bit by bit, but if you lose only a little information, you'll be able to reconstruct it closely enough so the human eye won't see much difference. If you lose a touch more, you may be able to tell the difference, but you might have to look for it. Lose a bit more, and the difference may be obvious, but you can still produce an acceptable image. And so on. Basically, you can trade off information for image quality, and choose between more compression with small file size and lower image quality, or better image quality with less compression and larger files.

JPEG Format

The standard format for compressing photos is *JPEG* (pronounced *jaypeg*), a lossy compression scheme named for the acronym for the Joint Photographic Experts Group. You can recognize files that use this format by the file extension .jpg (which is also pronounced *jaypeg*.). If you have compression turned on when taking a given photo, most, if not all, cameras will store the photo as a .jpg file. Some cameras don't use anything but JPEG format.

Lingo   *JPEG* is a lossy compression scheme that is the standard format for compressing photos.

JPEG is designed to let you choose between compression and small files on the one hand and maintaining image quality on the other, with a wide range of compression levels available. Individual cameras and programs can offer one or more compression choices. Most cameras offer at least two or three settings.

For an example of what JPEG compression actually does to a photo, take a look at Figures 3-16 through 3-19. The first picture in the group is not compressed.

Figure 3-16   An uncompressed image at 1200 pixels per inch width.

We took the photo in Figure 3-16 without compression, so it's an unadulterated 1200 pixels across. The original color version of this photo, cropped as shown, took 1.7 megabytes (MB) of disk space.

Figure 3-17   This is a compressed version of the photo in Figure 3-16.

The photo in Figure 3-17 is identical to the one in Figure 3-16, except that we compressed it using the maximum compression that our graphics editor allowed. Printed at this size, it's hard to see any difference between the uncompressed and compressed versions, and the compressed color version takes up only 70.8 kilobytes (KB) on disk.

Figure 3-18   Here's an enlargement of part of the uncompressed version of the file.

The difference between the compressed and uncompressed versions of the files shows up when you enlarge the image. Interestingly, the difference is far more obvious on screen than in printed output, but if you compare Figures 3-18 and 3-19, you can see the difference.

Note that in Figure 3-18 the enlargement is a little grainy, but there are no obvious problems in the photo. In Figure 3-19, you can see artifacts—problems introduced by the process of compression—in the form of little rectangular areas in many of the rocks, particularly on the upper left. These rectangles are much more obvious when you view this on screen, and they become more obvious in print with greater enlargements.

Figure 3-19   Here's an enlargement of the same part of the compressed version of the file.

Not so incidentally, we created the compressed version of the file by compressing the original, uncompressed version using Adobe Photoshop, so both started with the same image quality. In addition to comparing them for quality, look at Table 3-1, which compares the file size for different levels of compression for color versions of the image. To give a sense of the range available, we included file sizes for four different levels of compression, as well as the uncompressed file size. (To avoid confusion: we are not using Photoshop's names for the different compression levels. Photoshop calls them quality levels, so "low" to Photoshop means low quality with maximum compression. We are using "low" to mean low compression, and "highest" to mean maximum compression.)

Table 3-1   Sample of File Size with Different Levels of Compression

Compression Level	File Size
No compression (original .tif file)	1.7 MB
Low	367 KB
Medium	221 KB
High	142 KB
Highest	102 KB

As you can see in the table, the difference in file size between no compression and even the lowest level of compression can save a significant amount of storage space (1.7 MB compared to 367 KB). The difference between no compression and the highest level of compression can be enormous (1.7 MB compared to 102 KB). Note also that the file size for an *uncompressed* photo at a given resolution and color depth will always be the same. However, the file size of a compressed file will change from one file to another for any given compression setting, because the amount of compression you actually get depends on the amount of detail in the image.

The trade-off you're willing to make between compression and quality is very much a matter of personal taste, with no hard and fast rules to follow. You have to decide on the minimum quality level you're willing to accept based on what the pictures look like, and then weigh quality against the convenience of fitting more photos in your camera memory at once. But whatever quality level you're willing to accept, your choices are limited by what your camera offers—at least until after you move the pictures to your computer.

Now on to resolution.

About Resolution

Like compression, resolution has a visible effect on image quality. In Figures 3-16 through 3-19 we used the same resolution with and without compression. The next set of three pictures, Figures 3-20 through 3-22, are all without any compression, but with different levels of resolution. More precisely, the final versions of each photograph have the same number of pixels, because it's required by the printing process. However, all except the first started out with lower resolutions, and increasing the resolution for printing the photos doesn't increase sharpness or level of detail.

Figure 3-20   This first image looks sharp, with crisp, clean edges.

Figure 3-20 is a cropped version of the photo in Figure 3-16. In addition to cropping it, we then enlarged it to be 1200 pixels across. Even though we've enlarged the photo, it still shows crisp focus, which you can see along the edges of the rocks in the rock wall, and the leaves in the bushes above the wall.

Figure 3-21   This image, at 640 pixels across, has a soft focus effect compared to Figure 3-20.

In Figure 3-21, we took the original photo and dropped its resolution to 640 pixels across, which is a standard low resolution you'll find in many cameras. We then cropped it and enlarged it, following the same steps we used for Figure 3-20. Many people will still find this acceptable, but if you look closely you'll see a soft focus effect compared to the version in Figure 3-20. Not so incidentally, the color version of this file is 540 KB, compared to 1.6 MB for the color version of Figure 3-20.

Figure 3-22   At 320 pixels across this image goes beyond a soft focus effect to being outright blurry.

For Figure 3-22 we dropped the resolution down to 320 pixels across, a standard resolution for video clips. We otherwise followed the same steps as for Figure 3-21. You don't have to look closely at this picture to see that it's blurry.

Resolution and Compression Together

Both pixel resolution and compression work together to determine image quality. These two settings also determine how large a file the photo turns into, and how long it takes to save the photo to the camera's digital film. Saving a high-resolution photograph that's not compressed at all can take long enough to leave you drumming your fingers and watching the next good shot get away. And all of this taken together is why cameras use compression and why they offer more than one resolution setting.

As we already mentioned, choosing a compression level and resolution setting is very much dependent on what's available in your camera. That gets a little complicated to talk about, because cameras use different strategies for mixing the choices of resolution and compression settings.

Some cameras let you choose resolution separately from the compression, so you can combine any level of resolution with any level of compression. Others offer some resolution choices that are compressed and one or more choices that aren't compressed. Still others don't offer an uncompressed choice at all.

And you may or may not have different levels of compression to choose from for any given resolution.

Given all this, your first step in choosing resolution and compression settings is to find out what choices you have in your camera. Start by looking for information on resolution and compression, but be aware that your camera manual may use different terms. Among the cameras we gathered for writing this book, for example, the Nikon and Casio manuals talk about "image size" when they mean the resolution in pixels and "image quality" when they mean the level of compression.

Once you track down the settings available in your camera, you can decide which setting to use when. Given that the final choice is really a matter of taste, you'll ultimately come to your own conclusions based on your own experience. But there are a couple of tips we can give you to help you get started.

First, use the lowest resolution you need for any given picture. In Chapter 2, we pointed out that for viewing your photos on screen, you never need more pixels in the photo than there are pixels on the screen. More often, you'll want fewer pixels in the image, because you'll want to look at the image in a window. Even if the window is full screen, as in Figure 3-23, the title bar, menus, scroll bars, and so on will take up some pixels. If the photo has as many pixels as the screen, you either won't be able to see it all at once, as in the figure (which you can tell by looking at how much room the scroll bars give you for scrolling), or you'll have to zoom out to make it smaller, which may hurt picture quality.

Figure 3-23   For viewing an image on screen, you'll want fewer pixels in the image than in your screen resolution, or you won't see the whole picture at once at its best.

A useful rule of thumb for photos that you plan to show only on screen is to use a resolution that, after cropping, makes the photo roughly the same size as the next smaller standard for screen resolution. Table 3-2 shows the standard resolutions for PCs.

Table 3-2   **Photo Resolution for Different Screen Resolutions**

Screen Resolution (Name)	Screen Resolution (in Pixels)	Approximate Resolution to View Photos on Screen	File Size for Uncompressed Color Photo
Super VGA (SVGA)	800 × 600	640 × 480	0.92 MB
XGA	1024 × 786	800 × 600	1.4 MB
Super XGA (SXGA)	1280 × 1024	1024 × 768	2.4 MB
Ultra XGA (UXGA)	1600 × 1200	1280 × 1024	3.9 MB

Table 3-2 assumes that the lowest screen resolution worth taking into account is 800 × 600. Although there are certainly some people out there still using the older VGA (640 × 480) resolution, most computers today can handle at least SVGA (800 × 600), and those that can't are being used by people who aren't likely to look at photos very often. If you don't have a specific computer in mind for showing a photo—if you're putting it on a Web page, for example—assume viewers will have SVGA.

For printing, base your resolution on the largest size you ever expect to use for printing the photos. If you think you might print them at 8 × 10 inches, for example, you'll want to calculate the number of pixels you need to span 10 inches. If you want to print at 5 × 7 inches, you need only enough pixels for 7 inches.

The other number you need to know is how many pixels you want per inch in the printed photo. As we said in Chapter 2, 150 pixels per inch (ppi) produces reasonably high-quality output. You'll need to experiment with this a bit and decide whether 150 pixels is enough for your tastes. If you demand a somewhat higher quality, you may want to shoot for 200 ppi. Table 3-3 shows the number of pixels you need in the longer direction of the photo for some common photo sizes, for both 150 ppi and 200 ppi. If you expect to crop at all, keep in mind that each entry in the table shows the number of pixels you want *after* cropping the photo.

Table 3-3   **Photo Resolution for Printing at Different Sizes**

Photo Size (in Inches)	Number of Pixels in Longest Direction for Printing at 150 ppi	Number of Pixels in Longest Direction for Printing at 200 ppi
2.5 × 3.5 (wallet size)	525	700
4 × 6	900	1200
5 × 7	1050	1400
8 × 10	1500	2000

An important difference between picking resolutions for printing compared to picking resolutions for showing photos on screen is that for printing you can have more pixels than you need without hurting the quality of the printed photo. If you have a higher pixel resolution than you need, however, printing will take longer and you'll use up disk space unnecessarily.

Having said that, also note that taking a picture at a higher resolution than you need is not a problem for photos that you intend to view on screen. As we've mentioned elsewhere, you can lower the resolution with a graphics editor without losing image quality. We'll explain how in Chapter 8, "Fun with Pictures: Basic Editing." Right now, just be aware that it can be done.

There's one other thing you need to know about resolutions for printing: it's easy to get confused between picture resolution and printer resolution. We've been careful to talk about picture resolution in *pixels* per inch. Outside of this book, however, you'll often hear it referred to in *dots* per inch, which is a more common way to talk about resolution in general. And pixels are, after all, dots in some sense.

We're not splitting hairs by insisting on talking about pixels per inch instead of dots per inch; we're trying to avoid confusion. The problem is that printers measure printing resolution in dots per inch. If we used the same term to refer to pixels, and then we told you that you should print your photos at 150 dots per inch, you might think we were talking about the printer setting. But you want the printer set at some higher resolution—300 dots per inch, or 600 dots per inch, or higher still. No matter the printer resolution, however, you want the photo at 150 *pixels* per inch.

Finally, for choosing the resolution and compression to use, until you're familiar with your camera, take pictures with as little compression as the camera allows so you can learn how resolution affects image quality. Once you've sorted that out, you can start experimenting with the camera's compression settings or modes that use more compression to see what they do to image quality. In general, you should take pictures with the lowest resolution you need for the picture, and use compression only to the extent that you have to, either because the camera modes don't give you a choice, or it's the only way to fit a reasonable number of photos into a limited amount of memory in the camera.

What to Do with Your Photos After You Take Them

We almost named this section "Now That You've Got the Photos, What Are You Going to Do with Them?" With film the answer is easy: you take the roll to your friendly neighborhood drug store or photo shop, drop it off, and come back to pick up the pictures later. Then you do all the familiar things: mail them to friends, put them in photo albums, frame them, or put them in your wallet.

With digital photos, you have even more options, and some of them aren't so familiar. Much of the rest of this book is all about those options, so we won't go into any of them here in detail, but we can't finish a chapter on getting started with digital photography without at least mentioning what the options are—and where you'll find more information about them later in this book.

First, you'll probably want to move the photos to your computer. You'd be surprised at how many ways there are to do that (we cover them in Chapter 5, "Special Issues for Digital Photography," in the section "The Connection Choices: Cable, Docking Station, or Moving a Storage Card.") Briefly, your camera most likely came with a cable to connect to your computer—most often by USB (universal serial bus) port on recent models. Most cameras connect directly to the cable; a few use a docking station. Once you get physically connected, you may be able to simply open the photo files or copy them to your computer, as if the camera memory were just another disk drive. With some cameras and some versions of Microsoft Windows, however, you'll need to install a special program to move the files. You should find details on all of this in your camera's manual.

Once the photos are on your system, you may want to edit them to enhance the image, fix flaws, or turn them into greeting cards or postcards (all of which we cover in Chapter 8, "Fun with Pictures: Basic Editing," Chapter 9, "Advanced Editing: Fixing Flawed Photos," and Chapter 10, "More Fun with Pictures: Special Purpose Editing"); e-mail them or put them on a Web site (take a look at Chapter 13, "Sharing Your Photos: E-mail, Letters, and Web Sites"); or print them (Chapter 11, "Printing"). You can also upload them to a Web site to have the photos printed and delivered to your door by mail (we cover that one in Chapter 11).

If you have the right printer, you can print your photos without ever going near your computer. Just move the memory card from the camera to the printer, and print (see Chapter 11). Note too, that you if you think of printing the photos as a chore, you can get printed photos without printing them yourself, and without uploading them to a Web site. Some retail photo developing stores are equipped to download images from your camera's memory card and print them. And some locations, including drug stores, have kiosks where you can insert the memory card and pick the photos to print. (We cover this in Chapter 11 also.)

Our point, for now, is that you have a lot of choices. And some of them are, frankly, fun (at least we think so). So don't stop here: read on.

Key Points

◼ Check to see if your camera offers an adjustment to let you correct for farsightedness or nearsightedness. If it does, you can use the camera without glasses. Even if you don't need glasses, you may need to adjust the setting to see a sharp image.

◼ Make sure you know how to turn off the digital zoom or how to avoid using it if you can't turn it off.

◼ Make sure you know the closest you can get to your subject and still be in focus. If your camera has a macro mode, this will be a different distance with macro mode on than with macro mode off. Learn both.

◼ Watch out for differences between what you see in the viewfinder and what the camera sees. Find out what the differences are, and try to compensate when you frame the picture in the viewfinder. Use the LCD when framing really matters, which includes any time you use the macro feature.

◼ If you're using auto-focus or auto-exposure (or both), and you want to frame a scene with the subject off center, be sure to lock the focus and exposure with the subject in the center first by holding the shutter button halfway down. Then reframe the picture the way you want to take it.

◼ Some cameras offer clusters of settings that let you change several settings at once for common types of photos. This feature goes by different names in different cameras. If your camera offers this feature, learning the choices and what they are meant for is the easiest and fastest way to get good results from the camera.

◼ Use forced flash, also called fill-in flash, to fill in shadows when taking pictures in bright sunlight and to light a subject that is otherwise backlit—when you're taking pictures facing the sun, for example.

◼ Avoid using the red-eye reduction feature. The better choice is to raise the lights, tell your subject not to look directly at the camera, or fix the red eye later in a photo editing program.

■   Compression can be lossless or lossy. Lossy compression, which digital photos often use, loses information in the file. That translates into lower quality images, but much smaller files.

■   Don't use a higher resolution than you need for a given purpose. And until you are familiar with how resolution affects image quality, use as little compression as the camera allows. Then experiment with compression settings to see how compression affects quality.

Chapter 4

Is That a Snapshot in Your Camera, or Did You Take a Photograph?

In a book on photography, we'd be criminally irresponsible if we didn't talk about some basic photographic concepts—like composition, understanding the relative virtues of color versus black-and-white photography, and how to hold a camera steady. This is the chapter where we focus on the art and skill of photography from the point of view of the digital camera user. Be aware, however, that we're also about to take you a step further.

Most chapters in this book focus solely on practical, clearly useful tips and information. This chapter offers lots of practical advice as well, and you'll find that you can apply the practical tips to the real world right away. However, we also tread—mostly in the sidebars—into an area you might call "How to think like a photographer." Learning to take advantage of those suggestions and ideas will take much longer than learning the practical tips, but may ultimately do more to improve your photos than anything else in this book.

Many amateurs look at a good photographer's pictures with a feeling of awe at the power of the images, but they also think that given the time to take enough pictures, they could wind up with incredible photos, too. Don't believe it. Keep in mind the photographer whose detractors said that he was lucky to get such good photographs. "Yes," he said, "but I'm lucky so often."

Taking a good photograph doesn't require lots of picture taking, it requires that you aggressively look for a shot with an inquiring attitude, rather than passively accepting the first thing you notice. It means testing potential shot after potential shot in your mind's eye before you even aim the camera. It means opening yourself up to possibilities that may not be immediately obvious. It requires, in short, that you learn how to think like a photographer.

With just a little effort you can learn to change your vantage point, be more selective in what you want in the frame, and frame the image in a way that eliminates unnecessary elements. Learn how to do all this, and you'll have photographs instead of snapshots. But that's just the beginning.

The best photographers—both professional and amateur—have learned the art of seeing the scene differently than most people do, to spot the part of the scene that will turn the photograph into something more than it might otherwise be. That's harder to learn, but we can give you some hints in this chapter about how to do it.

So we define a photograph as art, as something special, and as something that took work and forethought rather than happenstance and good luck (or at least the hard-earned experience that lets you instantly recognize a scene as a photograph when happenstance and good luck hand it to you). And that's the difference between a photograph and a snapshot.

Let's start with some basics.

Basic Rules of Thumb for Taking Better Pictures

First things first: most rules are meant to be broken. We wanted to make that explicit, clear statement right up front. Keeping the color within the lines isn't a bad rule to start with for coloring books. But once you're confident that you know what you're doing, you can ignore the lines and even use different colors than the rules call for. Who says the sky has to be blue and the grass green?

The same can be said for many of the rules we give here. They're a good place to start, and you should make sure you have a feel for why they work visually before you break them with abandon. But don't get overly focused on following the rules. Otherwise you'll never learn how to break them and do something better.

That said, here are the rules—well, some rules, anyway.

About Snapshots   You might think from the title of this chapter that we're snobbish about photography, and that we turn up our noses at "mere" snapshots. So we'd like to put in a good word for snapshots.

Originally, the term *snapshot* referred to any photograph that stopped movement. In the early days of photography this simply wasn't possible. Camera equipment, glass plates, light-sensitive emulsion, and the like were slow, with exposures measured in minutes. Setting up the shot was a major undertaking, and whoever—or whatever—you were taking a picture of had to hold still for the camera. The camera had to hold still as well, which meant you absolutely had to use a tripod.

Taking a picture took time. Imagine seeing your toddler take his first step and wanting to take a picture with one of these early beasts. Quick: grab the tripod, set it up, attach the bulky camera, insert a glass plate, adjust the aperture and speed, put the black cloth over your head, and—while looking at the scene upside down and backwards and trying to compose the picture—snap the 3-second-long exposure. By this time your son has probably gone out on his first date. In short, by the time you saw something you wanted to take a picture of, it was already too late to take the picture.

The introduction of handheld cameras that could take a picture quickly changed photography forever. Suddenly, you could take a picture of just about anything anywhere, and whatever you were taking a picture of didn't have to hold still. The possibilities for photography were boundless. Rather than seeing life as a continuous flow of events that might occasionally stop and take time out for long enough to let you take a photo, you could now look at life as a series of photos you could take as is, freezing moments in time as they happened.

Digital photography has changed the nature of how we see things yet again by providing us with instant recall of what we've just seen. Like instant replay for a football game, digital cameras let you review the moment in time almost immediately. More than that, they let you stage-manage reality to get a better view. (Wives we know, who shall remain nameless, have been known to look at a picture and tell their husbands to take it again.)

Today, the quick and easy snapshot has developed a distinctive visual style. For the most part, snapshots are clean, unsophisticated, and even innocent in a way. And they are familiar. Most snapshots are reflections of what we do every day. That makes them emotionally powerful, because we can all identify with them.

Ultimately, there's nothing wrong with snapshots, but the picture you'd recognize as a snapshot is not the picture a professional photographer would take faced with the same scene. There are ways to take pictures that most people would recognize as good photographs, and any professional worthy of the name knows how. There's no reason that you shouldn't apply the same techniques to your own pictures.

Decide What You're Taking a Picture Of

This is so obvious that it almost seems silly to bother saying it: before you can take a picture, you need to know what you're taking a picture of. Of course you do; otherwise, how would you know where to point the camera, right?

Well, yes, but there's a little more to it than you might think at first. Take a look at Figure 4-1. Forget for a moment that this already exists as a photo and think of it as something you might see in someone's backyard.

Figure 4-1   A dog, a cat, and a trellis.

Presented with this scene and with a camera in hand, you may want to take a photo, but what are you taking a photo of? The cat on top of the trellis? The dog looking up? The interaction between the dog and the cat? Or do you want to wait for the dog and cat to get out of the scene so you can take a picture of the trellis and garden?

Each of these possibilities is a perfectly reasonable choice, and each would make a good photo. Before you can take any of them, however, you have to make a conscious decision about which one you're taking. You should be open to the possibility that your first thought for a photograph may not be the best choice. Actively look for other possibilities, and only then decide which picture to take.

Figure 4-1 is the photo of the interaction between the dog and the cat (or maybe lack of interaction; the dog is definitely watching the cat; we're not sure if the cat's interaction with the dog is pretty much limited to ignoring her). Figure 4-2 shows how the photo might look if you decided to take a picture of the cat, and Figure 4-3 shows the picture you might have taken of just the dog.

Figure 4-2   The same scene as in Figure 4-1, but using the cat as the subject.

Figure 4-3   The same scene again, but this time using the dog as the subject.

Figure 4-4, finally, is the picture you might have taken with the trellis and garden as the subject, after first waiting for the dog and cat to get out of the way as unwanted distractions.

As we said, all of these are perfectly reasonable pictures. The key point: you can't take any of them until you know what the picture is about.

Not so incidentally, you may notice that the photos of the cat and the dog have a lower resolution than the first picture in this set. That's because we created both by cropping the original photo. The resolution would have been higher if those had been the original photos we intended to take, instead of creating them as an afterthought by cropping in on a larger picture. And that brings us to the second rule: get in close.

Figure 4-4 The same scene yet again, without the dog or cat.

Get in Close

Thoreau once said, "You can't say more than you see." You may suspect he was talking about what you can say with words rather than with pictures, but it applies even better to photography, where it translates to a simple rule: Once you decide what you want to take a picture of, don't try to photograph it if you can't see it clearly. If you need to move in closer to see it—well, move in closer. It's that simple.

There is a whole different world when you're up close and personal. It's the difference between watching a basketball game from the thirtieth row of the second balcony and sitting at courtside. (Whether that's a good thing or not depends on whether you're a sports fan, but it's a different experience either way.) Sitting courtside, you're immersed in the experience; you get details you would never see from a distance. You hear the grunts and groans of the players, see the sweat pouring off their faces, and hear their sneakers squeak as they turn on the hardwood floor.

More important, when you're up close, the game fills your entire field of view, both literally and by way of a kind of psychological tunnel vision. You have fewer distractions and get to focus more on the game.

Photos that close in on the subject of the photo do much the same thing. As we've suggested in earlier chapters, they focus the viewer's attention on what you want him or her to see, and they eliminate distractions.

Having said that, however, we need to make the distinction between being physically close to the action and getting a close-up courtesy of a telephoto lens. If the alternative is to get a shot that isn't close up in any sense, then using a telephoto lens is much better than not using one. It gives you most of the benefits of being right there, and will let you get candid shots without distracting your subject. But there's something to be gained from being physically close. And if you really want a photo that feels fully immersed in the action, you need to go in close yourself.

Telephoto Isn't the Same as Being There

Now, obviously, if you want a picture of a speedboat making waves, you won't (usually) put on a pair of water wings and swim to the nearest buoy. Instead, you'll use a telephoto lens. Similarly, if you're trying to get a picture of someone who instinctively poses for the camera when you're there, and you can't get him or her to relax to get a candid shot, try sneaking in close by way of a telephoto lens. But for a picture of your child first learning to walk, or your first new car, don't use a long lens to take the shot. Move in close.

One issue is that you get distortion from a long focal length lens. Foreground and background are visually compressed in space, so objects in the scene appear closer together front to back than they actually are (which is fine if that's the effect you're trying for, but not otherwise). More important, when you shoot from a distance, you are not in contact with the action, and that makes an even bigger difference.

Getting in close opens up all kinds of possibilities you wouldn't have seen or thought of, because you won't notice them except on close inspection. (For an example, take a look at the sidebar "Seeing at Different Distances.") But there's another issue too: sometimes it's not so easy to allow yourself to get in close.

Try This! With camera in hand, turn around 180 degrees and pick some object—any object—that's at least a few feet away that you'd like to photograph. Then take two steps closer and look at it again. Then two steps more and take another look. Notice the shape and form and color. Odds are that you'll see some details—and often a more interesting possibility to photograph—that you didn't see from your original position. That's the advantage of being close up yourself. The closer look, which you won't get the same way from using a telephoto lens, may change the way you see the object and change your thinking about how you want to photograph it.

The next time you're about to take a picture of someone, take one step forward. You'll be surprised how hard it is to bring yourself to do it. Then give your camera to a friend and ask him or her to take your picture. More often than not the first thing he or she will do is take a step backwards. Wrong move.

If your photograph is about the relationship between one subject in the picture and another (like the dog and cat in Figure 4-1 that we discussed earlier), then, by all means, step back to get it all in. More often than not, however, it's a picture of a specific person or object. To make sure that the subject dominates the photo, you may need to step forward.

The problem is that most of us are not comfortable invading other people's zones of privacy, so we stay at a distance. As a photographer, you need to get beyond that (preferably not too far beyond, or you'll turn into an obnoxious *paparazzo*—that's one of the *paparazzi*). We repeat: by getting close, you give yourself a better opportunity to explore your subject and discover significant details you wouldn't see otherwise. And that can only make your picture better.

Seeing at Different Distances

Here's an example of how your thinking about a potential photograph may change as you get in close to a potential subject. Since this is very much a personal experience, we'll tell it in the first person, as it happened to one of us.

I once wanted to take a photograph of a garden that happened to have a statue in it, just because it was pretty. Figure 4-5 shows the scene.

Figure 4-5 A view like this might catch your eye.

As I got in closer, I began to focus on the statue, first in its relationship to the tree behind it and the landscape around it. Then, as I got closer still, I began to focus on the statue itself. The surface had pitting holes from aging, and the statue began to come alive for me. The final result was a very different photograph than the image I had originally seen of a statue on the lawn, as you can see in the photo at the top of Figure 4-6 and again (at even closer range) in the photo on the bottom.

Figure 4-6 These are two very different photos of the same statue that's in Figure 4-5.

Anticipate the Action

Except when you're carefully setting up a shot of things that don't move—like a taxidermy collection—photos are invariably targets of opportunity. The universe at large, and your intended subject in particular, is not always going to stop while you get ready to take the picture. What happens, quite simply, happens. It's your job to be ready for it.

The good news is that there are times when you can pretty well predict what's likely to happen. And if you take the time to make that prediction, you have a better chance of getting the shot.

Predicting is not so hard. When your husband receives an award at a banquet, he'll probably shake hands with the person giving it to him. Be ready for it. He's going to walk up to the presenter, take the award with his left hand and, unless he's really short on social skills, you know his right hand will reach out to shake hands. *Click*; there's your shot.

When you're trying to get a photograph, you have to focus on the photograph, not on the things that you might otherwise pay attention to, like the nice things the presenter is saying about your husband. While everyone else is applauding, you need to be predicting the timing: when will he get to the podium and how long will it be before you can get the photo?

When the action happens, shoot a bit earlier than the moment you want to freeze in time. It takes a few milliseconds for your brain to recognize you've got a shot and tell your finger to push the button, and a few milliseconds more for the shutter on the camera to open and shoot the picture. You need to compensate for that. A little practice will give you a feel for how much earlier to shoot.

Try This! You can easily practice anticipating action. Go to a place where there's action to shoot, like a Little League game or a karate school. Don't worry whether you know anyone there or not, as long as they'll let you take pictures. Get as close to the action as you can and take pictures of the batter swinging at a pitch, or the black belts doing their thing. See if you can capture the exact moment when the bat hits the ball, or a hand or foot technique comes to its final point. Go a little crazy, and take lots of pictures. Remember, you're practicing, and you're using a digital camera, which means you can take as many pictures as you like and then erase them at no cost (except for the price of batteries, which are a lot cheaper than film and developing). After each shot, check the image in the LCD and see if you caught the shot the way you wanted it.

When you've finally succeeded (and you will), practice some more. The point of this exercise is to make you aware of the process that lets you get the shot. It's too much to expect that you'll stop the action at just the right point every time, but the more you practice, the more often you'll succeed, and get a shot like the one in Figure 4-7.

Figure 4-7 Catching action as it happens—in this case as the wood breaks—takes practice.

One other thing: in Chapter 3, "Getting Started with Digital Photography," we urged you to turn off that annoying red-eye reduction feature on your camera. When you're taking an action shot it's absolutely essential, so the camera takes the shot when you squeeze the shutter button and not a couple of seconds later. If you're taking the shot for real, and red-eye reduction is on when the black belt is about to break the wood, you'll still get a picture, all right, but you won't catch the moment when the wood breaks. You'll have a picture of people walking away, holding pieces of shattered wood.

Anticipate the Shot

In addition to anticipating the action, you'll want to anticipate the shot, which is a related, but slightly different issue. Anticipating the action means thinking about *when* something will happen. Anticipating the shot means thinking about *how* you want the picture to look when it happens.

The famous French photographer Henri Cartier-Bresson would find a wonderful frame of a picture where all the pictorial elements worked together. He'd then wait endlessly, like a cop on a stakeout, for a particular action to happen within the scene he'd framed. He already saw the end result in his mind's eye, which made the final result worth the wait.

You probably won't ever wait as long as Bresson did, because he was a photographer and that was his life. But you can learn how to frame a picture in your mind, and in the camera, so you can be there to capture it. (For an example of what you may have to go through, take a look at the sidebar "Waiting for the Shot.")

Say that you're sitting in a restaurant when a window washer starts cleaning a window near your table. You realize in your mind's eye that it would make a good photo if you could get a picture of him in just the right position with his arm extended and water smeared across the window, distorting the view through the glass. You pick up your camera, frame the shot, and wait. If it all works out, he'll follow the script and, *click*, there's your photograph. You'll find it in Figure 4-8.

Figure 4-8   A window washer, framed in the window.

Waiting for the Shot
Here's another example of how you might think about a photograph, in this case dealing with a picture you know you want, and are waiting to happen. Here again, since this is a personal experience, we'll tell it in the first person, as it happened to one of us.

Just before starting this book, I moved into a new house in an area that used to be serious horse country, and still has plenty of horses. Literally across the street is a 200-plus-year-old farmhouse, a barn, and a fenced-in field, which often has two horses in it.

On a bad day, the view stepping out the front door is great. On a good day, it's picture perfect: sunny, blue sky, puffy white clouds, green grass, red barn roof, white farmhouse, wooden fence—you get the idea. From the day I moved in, I knew that I wasn't going to send people e-mail with a picture of my new house. I was going to send them a picture of the view out the front door.

That was two months ago as I write this. I still haven't gotten the shot.

First, I had to find my camera (no small feat given the number of boxes that moved with us).

Then I needed a perfect day with the right weather. (No blue sky, no puffy clouds, no picture.)

Then the people who own the farmhouse decided they were moving, and the people who kept their horses there decided they needed to find a different place. (No horses, no picture.)

Then some other people decided to keep their horses in the field, but only overnight, so they walk them away every day at about 9 a.m. That means the shot I want is possible again, but only from about 7 a.m. to 9 a.m.

So I wake up every morning and look. One day I walked out at about 8:45 a.m. and it was just right. Blue sky, puffy clouds, two horses grazing in the field. I ran inside to get the camera, ran back outside, and...there was my wife, who happened to be walking our dog, standing in the shot. I called to her, and asked her to move, but she had to move a lot to get out of the shot. When she finally made it, the horses were gone; the woman who keeps them there had come to get them for the day.

So I didn't get the shot that day either. But I will. I have the photograph perfectly composed and ready to take. In fact, I already have the photograph done. I'm just waiting for reality to catch up to me when I have a camera in hand. Then you'll be able to see the photo too.

Quick Rules

Some rules of thumb take just a few sentences to explain (which is why we gathered them together here, under one heading), but that doesn't make them less important. In fact some of the most important rules are the short ones, starting with holding the camera steady.

Hold the Camera Steady

We've always heard photographers say to their subjects, "Hold still," but we've never heard anyone say that to photographers. They should, because one of the surest ways to ruin a good photograph is to shake the camera. You'll shake it with your natural body movement, or by pushing the shutter button like you're stabbing at an elevator button and being impatient about it. So the first quick rule is to be concerned about holding the camera steady. Be very concerned. The next several rules make some suggestions about how you can do that.

Got Image Stabilization?

It helps if your camera has built-in image stabilization. If you're lucky enough to have a camera that includes this feature, read the manual to learn how to take best advantage of it.

Be a Tripod

There are also ways to hold the camera steadier. Hold your elbows tight to the trunk of your body or rest them on a flat surface. Let your hands and camera rest against your face. You can also lean against a wall for added stability. Try not to breathe while you're taking the picture, and use a gentle squeeze on the shutter button instead of a sudden push.

Get a Tripod (or a Monopod)

Get and use a tripod or a monopod (basically a stick mounted on the bottom of the camera). Monopods are a good compromise for holding the camera steady while still letting you move around quickly—at a sporting event, for example. It's easy to rest the monopod on the ground or some convenient surface, then pick up the camera and monopod to move to another spot and get ready for the next shot.

Use the Timer Feature on Close-In Shots

You can avoid the camera shake associated with pushing the shutter button by using the camera's timer feature, if it has one. This works particularly well with macro shots, where you're in close, it's hard to stay in focus, and the force of your finger on the shutter can move the camera. Simply set the timer, frame the picture, and wait for the timer to go off.

Make Sure the Shutter Speed Is Fast Enough for Handheld Shots

If your camera lets you control shutter speed with a manual mode or shutter speed priority mode, make sure the shutter speed is fast enough for the lens you're using. The rule is that the minimum shutter speed for handheld photos is one divided by the focal length of the lens—or more precisely, one divided by the equivalent focal length for a 35mm camera. If you're using the equivalent of a 50mm lens, don't set the shutter speed to anything slower than 1/50th of a second. With a 150mm-equivalent lens, don't use a shutter speed slower than 1/150th of a second, and so on. (Just to avoid confusion: lower numbers in the denominator indicate slower speeds; 1/50th of a second is slower than 1/150th.)

Make Sure the Shutter Speed Can Stop the Motion

The rule about speed and focal length is for standard photographs, without anything moving except you as a camera platform. If you're shooting something that's moving and you want to freeze action, you'll need a shutter speed that's at least four times as fast as for standard shots, meaning that it stays open one fourth of the time or less. That translates to 1/200th to 1/400th of a second for the 50mm-equivalent lens, 1/600th to 1/1200 of a second for the 150mm-equivalent lens, and so on.

Setting Depth of Field

If you're taking a picture and you care about depth of field—either to make sure everything is in focus, or to make sure it's not—you'll need to set the aperture by changing the *F-stop*. It may help to know that the higher the F-stop number, the greater the depth of field. So an aperture setting of F-8 will give you greater depth of field than F-2.

Lingo   *F-stops* are the aperture settings for a camera lens.

Controlling Exposure and Why You Might Want To   With automatic exposure, you don't have to know anything about exposure settings. But knowing can be useful. If you understand what the camera is doing to expose your picture correctly, there are ways to take advantage of that knowledge.

As we discussed in Chapter 2, "Knowing (and Choosing) Your Camera" in the section "Choosing a Level of Control," the shutter and the aperture together control the amount of light getting into the camera. The camera figures out what that exposure should be and then adjusts both settings. However, the settings the camera chooses aren't the only ones that will give you the right amount of light, and they're not necessarily the settings you want to use.

If your camera has an aperture priority mode, you can set the aperture so you can control depth of field and let the camera figure out the shutter speed. If it has a shutter speed priority mode, you can set the shutter speed to stop action or create a blur, and let the camera figure out the aperture.

Unfortunately, a fair number of digital cameras offer a full manual mode and a full automatic mode, without an aperture priority or shutter speed priority. The good news is that in those cases you can still set the aperture or shutter speed the way you want it if you learn how to use the manual mode. We suggested in Chapter 2 that full manual mode is best left to prosumers and professionals. But having said that, we still think you should know how to do it if you want to.

Here's how: Standard shutter speeds are designed so each step to a faster speed will cut the amount of light getting into the camera in half. Similarly, standard aperture settings—the F-stops—are also designed so each step of smaller aperture will cut the amount of light in half. The usual F-stops (if your camera lets you set them) are 2.8, 4, 5.6, 8, 11, and 16. And—here's the secret—if you move one of these settings in one direction, to cut the amount of light in half, you can exactly compensate for that loss of light by moving the other setting in the other direction, to double the amount of light.

Given this relationship between shutter speeds and F-stops, once you know one pair of settings that gives the right amount of light, you can figure out the other pairs of settings that will let the same amount of light into the camera. Simply adjust shutter speeds in one direction, to let more or less light in, and F-stops in the other.

For example, suppose you're in fully automatic mode and your camera comes up with a setting of F5.6 at 1/125th of a second. You know you can get the same amount of light by moving the shutter speed some number of steps faster and opening the aperture the same number of steps wider (toward a lower number). Or, you can get the same amount of light by moving the shutter speed some number of steps slower, and closing down the aperture (moving to higher numbers) by the same number of steps. If you can remember that simple rule, you can use manual mode. Let the camera find the automatic settings, then choose the shutter speed or aperture you want—depending on whether you care more about depth of field or stopping action for the particular shot—and adjust the other setting by the same number of shutter speed settings or F-stops. And that puts you in full control.

Line Up with the Horizon

Here's a rule that sounds so simple that it seems hardly worth mentioning. But it's an important consideration for adding structure to your photographs, and it's amazing how often the rule gets ignored. You generally want to line up the horizon—or whatever serves visually as the horizon in a given picture—so it's parallel with the edge of the picture. This is one of those little things that doesn't stand out when done right, but becomes a glaring annoyance when it's done wrong.

For example, Figure 4-9 is a simple, clean photograph of a flower with a tree line that serves as a visual horizon behind it. It's not a great photograph, but it qualifies as pleasing.

Figure 4-9   The tree line, which serves as a visual horizon, is parallel to the edge of the picture.

The version in Figure 4-10, however, shows the horizon skewed in relation to the frame. Now, it happens that the camera is being held parallel to level ground and the visual horizon is actually a hill. But in the absence of any other visual cues that might suggest that it's a hill, this mismatch between edge and horizon makes it look like you took the picture without bothering to frame it in the viewfinder or LCD.

Figure 4-10   When the horizon isn't parallel to the edge of the picture, the photo can look like it was snapped by mistake.

Line Up with Vertical Elements

Closely aligned with the use of the horizon is the use of vertical elements, which can also help with framing. The picture in Figure 4-11, for example, looks odd, because the bird feeder is on an angle to the left and right edges of the frame, despite the fact that the tree line is parallel to the top and bottom edges. But your experience with the real world says that the feeder should be hanging straight down. The problem is that the tree line is actually angled up a hill.

Take the picture with the bird feeder parallel to the edge of the frame, as in Figure 4-12, and the photo looks fine. Given the bird feeder as a visual clue, it doesn't look odd that the tree line, as a visual horizon, isn't parallel to the bottom edge of the photo.

Choosing a Composition

Very often, you'll look at photographs and be attracted to them. Sometimes it's the subject matter that you're drawn to, but sometimes it's something you can't put your finger on. There are always things going on in a photograph that don't register on a conscious level but make the photos come alive. Or make them fall flat. More often than not, these issues fall into the general category of composition. Composition is a massive subject worth a book of its own, and we won't try to tackle it in any depth here. But we will point to one simple rule that can often improve your pictures if you keep it in mind: the rule of thirds.

Figure 4-11   The tree line is parallel to the edge, but the bird feeder is at an angle, which doesn't match the real world.

Figure 4-12   Line up the bird feeder with the edge, and the picture looks right.

Rule of Thirds

When composing a photograph, it feels comfortable to put the subject of the picture in the center. And often that's the right place for it. The subject is the center of attention. Besides, isn't that where the viewfinder sensors for auto-focus and auto-exposure are located? Well, yes they are, but that doesn't mean you have to center every picture you take.

If you depend too much on putting things in the center, some photos just won't work visually. For example, Figure 4-13 shows an outdoor scene that's centered on the horizon. As a viewer, on a level so deep that you may not even be aware of it, you don't know what you're supposed to focus on in the photo. What's more important, the part below the horizon or the part above it? Should you be looking at the ground or the sky? The result is that overall, it's a boring, unsatisfactory photo.

So if you're taking the picture of a horizon, where should you put the horizon in the frame? Often, the answer isn't immediately obvious.

The rule of thirds offers one possibility. It says that you can break up your frame into thirds, both horizontally and vertically, and then place horizontal and vertical elements along any of the lines, place the subject at any of the points where the lines intersect, or both. This avoids an overly centered look and often produces interesting photos. As one example, Figure 4-14 is similar in many ways to Figure 4-13, in that it also includes a horizon, but the horizon is about one-third of the way down from the top of the photo, which is a visual cue telling you to look at the ground rather than the sky. As a bonus, there's a giraffe standing about one-third of the way in from the right side. In this photo, at least, the rule of thirds rules.

Figure 4-13   Centering a horizon usually produces a boring photo.

Figure 4-14   Using the rule of thirds, you'll often wind up with a more interesting photo.

Watch Out for Unwanted Elements

We almost called this section "You Can See a Lot Just by Looking," which is cribbed from Yogi Berra's "You can observe a lot just by watching." Our point (which is probably different from Yogi's) is that there's a difference between what you see and what the camera sees, and you have to learn how to bridge that gap. Once you learn to do that, you'll be able to see those unwanted elements we mention in the section title, which is the first step in getting rid of them.

As a member of the human race, you're subjective. You tend to see things as you want them to be, not as they are. You tend to concentrate on what Gestalt psychologists refer to as *the figure*. Everything else is the ground, or background.

The figure can change depending on what is important to you at the moment. For example, the book you're reading is (we hope) currently your foreground. (We're about to change that.) The feel of the chair on your buttocks, the lamp that's providing light, and the sound of the air conditioner (or the crackling fire) are background. Or they were background. Now that we've drawn your attention to them, you're probably feeling the chair below you and hearing that air conditioner or the crackling of the fire. The point is that the foreground, or figure, can and does change all the time, depending on what you're paying attention to.

Unlike our visual system, the camera lens is indiscriminate. That's the essential truth behind the old saying that the camera doesn't lie. Cameras don't suffer from the psychological tunnel vision that we just described, and they don't care what's important to you. They'll take a picture of what is actually there, not just the part you're concentrating on. Unless you control what the camera sees, the photo will not reflect your interests or sensibilities, which is to say, it won't take the photos you mean to take when you're looking at a particular scene.

Figure 4-15 shows a shot of rustic old buildings that is pretty well ruined by a combination of telephone poles, garbage cans, and telephone wires. It's not unusual for people to take a shot like this, without noticing the clutter until they see the picture.

Figure 4-15   Telephone poles, garbage cans, and other clutter may not stand out when you're taking a picture, but they show up in the photo nevertheless.

The trick to avoiding unpleasant surprises like this is to train yourself to see what's actually in the frame when you look through the viewfinder or at the LCD. Get past the scene your tunnel vision is showing you and see what's actually there. Once you can do that, you can learn how to compose the picture based on how it will actually look.

Learning How to See the Shot Here's an exercise that can help you learn to see like a camera (or a photographer): Pick out a subject to photograph and frame the shot. Now look in the upper right corner of the viewfinder or LCD. What you see, remember, will be in your picture. Consider whether you want it there. Then do the same with the other three corners and with your subject. Look for distracting objects sticking out from behind the subject and things in front as well. Keep reminding yourself that anything you see will be in the picture. And consider whether you want it there. Try this exercise often enough, and it will become automatic. You'll learn how to see like a camera, which is to say, like a photographer.

Once you start looking at a scene this way, you can improve your photos by physically moving unwanted objects out of the way, or by positioning yourself, the subject, or both so the unwanted elements don't show. Figure 4-16 shows the same rustic buildings as Figure 4-15, but in this case, the picture shows what you want people to see when they look at your photo.

Figure 4-16 You can improve a shot by moving your position to eliminate clutter.

Black and White Versus Color

Many photographers will tell you that one of the surest ways to ruin a perfectly good photograph is to put color into it. While you may not feel that way, the virtues of black-and-white photography are worth considering.

As a digital photographer, you have one important advantage over film photographers: You don't have to commit to black and white or color when you buy film. You can switch between the two modes at the press of a button, or easily convert color pictures to black and white later. That means you can choose which one to use even as you form the picture in your mind's eye. If you're like most people, however, you'd never think to use black and white if color is an option. So the question is: why use black and white at all?

In many cases, black and white is the most effective way to bring out the point of the photo; if you're trying to focus on the compositional or emotional elements in a scene, color is window dressing that gets in the way. For example, it's more difficult to look at a photo and feel emotion for a crying child who is wearing a bright yellow shirt than it is for the same child if the picture is in black and white, and the shirt is a shade of gray. Or think of the most gripping news photos you've seen—the ones that win Pulitzer prizes and burn themselves into your memory. What you remember from each photo—what *we* remember at least—is the expression on a face, the composition, the raw emotion. Color would be beside the point, and could even be distracting. Black and white strips the image to its bare essentials.

As one example, consider Figure 4-17, which is a close-up of part of the front of a fire truck. For many viewers, this combination of fire truck and American flag will resonate with images of Fourth of July parades in Smalltown USA, and fire departments that still show up to get cats out of trees. However, the photograph is also about more than a fire truck. The photographer's interests lie in several areas at once. The large textured grill on the right, combined with the visual elements of circles, a bell shape, and an American flag on the left, gives a sense of power and softness working together. The power is poised and waiting for action, perhaps to rescue the next cat stuck in a tree.

You may disagree, but we submit that this photo would not be as interesting in color. More generally, if the point of a photo is its compositional elements such as shape, form, or contrast, or if the photo is about making an emotional connection, black and white will often do the job better than color.

Figure 4-17 For our tastes, this photo is more compelling in black and white than it would be in color.

And that, quite simply, is why you should consider using black and white.

One Last Thing

We'll end this chapter with one last thought. Bresson, who we mentioned earlier as being known for staking out a scene and waiting for just the right shot, once said that making a photograph was like the last three words in James Joyce's *Ulysses*: "Yes, Yes, Yes." Any photographs that didn't rise to that level, he said, should go in the trash. He meant trash literally, which can get expensive when you're paying for film and developing. But as a digital photographer, your trash is a simple delete button, which makes it a lot cheaper to trash the photos that scream "Almost" instead of "Yes." So don't be afraid to try to capture the exact right moment, and don't hesitate to throw away the near misses.

Key Points

- Think about what you're taking a picture of: is your subject one person, or the interaction between the one person and another? Then get close in to the subject, whatever it is.

- Getting close with a telephoto lens isn't the same as actually getting in close, where you might perceive the scene differently. Get physically close yourself when you can; reserve your telephoto lens for when you can't get close for some reason, or when getting close might distract your subject and ruin the shot.

- Try to predict the action, and be ready to snap the shutter when it happens.

- Try to envision the actual shot. Know the shot you want to take, and then wait for reality to accommodate you.

- The single most important rule for photography is to hold the camera steady. Get a tripod if necessary.

- Make sure the shutter speed is fast enough both to compensate for your own body motion when you're holding the camera, and to stop the action if that's what you want to do.

- Higher F-stop numbers give greater depth of field than lower F-stops.

- Line up with the horizon, or line up with a vertical line, like the side of a building.

- When composing a photograph, consider whether it will be a more interesting shot if the subject is off center. The rule of thirds says that you'll often get an interesting view if you divide the scene in thirds both horizontally and vertically, and place the subject along one of the imaginary lines or at one of the points where the lines cross.

- Watch out for unwanted elements, like a tree branch sticking out from your subject's ear. When you're framing a shot, the tendency is to fall into a kind of psychological tunnel vision and concentrate on what you want to take a picture of. Consciously practice forcing yourself to look at what else is in view, and then decide whether you want it there.

Chapter 5

Special Issues for Digital Photography

We've known people who go on a vacation, take lots of pictures on a film camera, come home, put the rolls of film in a drawer, and don't get around to developing the film for two or three years. (This is not a good idea, by the way. Film goes stale, which, among other issues, changes the color quality of the pictures you get.) We don't understand why people would do that, but some do.

Most people like to see their pictures a lot sooner. With digital cameras, that means printing the pictures or viewing them on screen. Either option usually (but not always) means moving the files to your computer. So you have to learn how to do that. Also, if you print the pictures yourself, you're well advised to learn a little about the printer you're using, and the possibility of getting a different one that might give you better results—something film photographers never have to consider (except, maybe, in the very broad sense of picking a better place to have their film developed).

Then there's the question of how to best store your photos.

Hard disk capacities have grown so large, so quickly that if you've bought a computer or upgraded your hard disk recently, you may feel you could never imagine filling it. (What will you ever do with a 60-GB hard disk?) If you take many pictures, however, you'll find that there is no such thing as a hard disk that's big enough.

You'll also find that throwing all your pictures onto your hard disk without organizing them isn't quite the same as throwing them all in a shoebox for sorting through later. For one thing, it's not as much fun to browse through files on your hard disk. For another, having your photos in digital form offers lots of possibilities for more efficient storage—not to mention the ability to print out a new copy of your photos at any time without worrying about fading or otherwise aging negatives.

All of these issues are different for digital photography than for film photography, and all are important to understand if you want to get the most out of your digital camera.

Let's start with choices for moving the photos from your camera to your computer.

Getting the Photos Out of Your Camera

Your camera's manuals should tell you everything you need to know about moving files to your computer. But they may not tell you everything you ought to know. There are two basic issues involved in moving photos to your system. One is how you make the physical connection. The other is what software you use to move the files. With most cameras, you have at least two choices for each of these categories. Most manuals will tell you about only one of them. We'll try to fill in the gaps, which may include sketching out some options you didn't know you had.

The Connection Choices: Cable, Docking Station, or Moving a Storage Card

Before you can move your photos to your computer, you have to somehow make a connection between the memory card and computer. One way to do that—and the only way most camera manuals mention—is to leave the storage card in the camera, and then connect the camera to the computer. You should have gotten a cable for that purpose along with the camera. The cable plugs into the camera on one side and the computer on the other.

The type of cable depends partly on the camera manufacturer and partly on how old the camera model is. If it's an older model that you bought years ago, the computer side probably ends in a serial connector, like the one on the left side of Figure 5-1. If it's a newer model, it probably uses a universal serial bus (USB) connector, like the one on the right side of the figure. The camera side of the connection doesn't matter for our purposes, as long as it plugs into and works with your camera. You should be aware, however, that the connection varies depending on the camera. One of the more interesting possibilities, which Kodak uses, is to end in a docking station on the camera side. Instead of plugging the cable into the camera, you simply put the camera in the docking station.

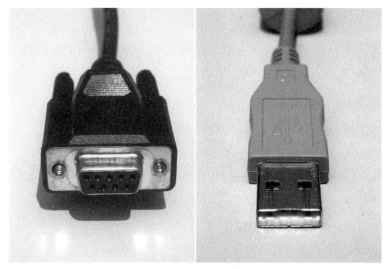

Figure 5-1   Older cameras most often use serial cables, as on the left. Most newer cameras use USB cables, as on the right.

All of these choices are variations on a theme. And which one you use depends entirely on the camera you have. However, there is another option that most camera manuals don't mention: you can remove the memory card from the camera and plug it into a slot so the computer can read it directly.

There are at least three reasons you might prefer to move the memory card to your computer rather than plug in a cable between the camera and the computer:

■   First, with most cameras, if you connect to the camera by cable without also plugging in a power cord, you'll be using up battery power as you look though the photos and move them to your computer. (A few cameras actually charge the batteries while they are connected.)

■   Second, there's no good option for dealing with the cable. If you leave it plugged into your computer all the time, the camera end will hang loose and add to the clutter on your desk. If you unplug it when you're not using it, you'll have to find a place to store it, then retrieve it and plug it back in when you need it.

■   Third, if you have more than one gadget that plugs into your computer—a handheld computer, a digital recorder, an MP3 player, a second digital camera, or whatnot—the cable clutter will be that much
worse, and anything you can do to minimize it is a good idea.

Alas, plugging the memory card into the computer isn't quite as easy as it
sounds, but it's not all that hard either. It's not as easy as it could be, because
computers do not generally come with the right slots for memory cards. (This is
meant as understatement.) It's not all that hard, because this oversight is easily
fixed. One straightforward way around the problem is to add the appropriate
card reader to your computer, which can be as easy as plugging in a cable. Figure 5-2, for example, shows a SanDisk card reader that works with Compact-
Flash cards and SmartMedia cards.

Figure 5-2   Card readers like this one can attach to your computer so you can simply insert your camera's
memory card and move the files to your computer.

There are more variations on card readers than we can conveniently list here.
For every type of memory card, there's a reader. More than that, there are readers
for most combinations of memory cards. The SanDisk card reader in Figure 5-2,
for example, reads the two most common memory card formats. You can also find
readers that can read every memory card format currently available.

Obviously, if all you're interested in is a reader for the card in your camera, the reader doesn't have to read anything but that one format. However, there are good reasons to have a multi-talented reader instead. Suppose you have—or eventually wind up with—a camera that uses CompactFlash cards, a digital voice recorder that uses SmartMedia cards, an MP3 player that uses Memory Sticks, and a *personal digital assistant (PDA)*, or handheld computer, that uses MultiMediaCards.

Lingo   *Personal digital assistant (PDA)* is another name for a handheld computer.

Instead of getting a separate reader for each one, it would be convenient to have a single reader on your desk that can handle all of these formats. So before you buy a reader, consider all the formats you may eventually need. Or play it safe, and get one that reads all the formats currently available (and hope no one comes up with a new format before you get your next digital gadget).

Before you get a reader, you also need to know that there are variations on how the reader attaches to the computer. Most current models connect to the USB port. That's generally the preferred choice, as long as your computer has a USB port and you're using an operating system that supports USB. If you have any doubts about your system, take a look at the sidebars, "Does Your Version of Windows Support USB?" and "Does Your System Offer USB?" Also, before buying a reader, check to make sure it works with your version of Microsoft Windows.

If your computer or your version of Windows predates the USB port, you need a reader that connects in a different way. The most common alternative is the parallel port, which is widely known as the printer port, since it was the standard connection for printers before USB took over that role. Your computer almost certainly has a parallel port, but it may be taken up with a printer. Either make sure that there's a second port available for the card reader, or get a card reader that includes a pass-through port. This gives you a place to plug in your printer and let the printer information literally pass through the card reader. The pass-through feature lets you plug the card reader into your parallel port and plug the printer into the card reader. Here again, before buying the reader, check to make sure it works with your version of Windows.

Does Your Version of Windows Support USB? To connect a USB reader, you need a version of Windows that supports USB. If you have Microsoft Windows 98 Second Edition (SE), Windows Me, Windows 2000, or Windows XP, you have solid Windows support for USB. Otherwise you don't.

If you have Windows 98, and you're not sure if it is Windows 98 SE, right-click the My Computer icon on the desktop and select Properties to open the System Properties Window. You'll see a screen like the one in Figure 5-3 that will identify the version as Windows 98 (as in the figure) or as Windows 98 Second Edition.

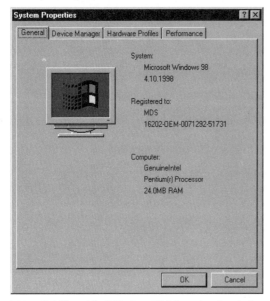

Figure 5-3 The original Windows 98 (shown here) and all versions of Microsoft Windows 95 have limitations on their USB support, or no support at all.

If you have the original Windows 98 or any version of Windows 95, USB support is problematic. Microsoft introduced USB support with Windows 95 OEM Service Release 2.1 (OSR 2.1). If you have an earlier version of Windows 95, you don't have USB support at all. If you have OSR 2.1 or later, you have USB support, but there are known problems with some USB devices under some circumstances. The original version of Windows 98 does better, but not as well as later versions, starting with Windows 98 SE.

To play it safe, we'd advise staying away from USB devices unless you have Windows 98 SE or a later version. If you're brave enough to try using USB with earlier versions of Windows, be aware that you may run into problems.

Does Your System Offer USB? Assuming you have an appropriate version of Windows, you also need a USB port. If you have an older system, it may not include one. Look for a rectangular connector like the ones in Figure 5-4. The ports may be on the front of the computer, the back, or both. If you can't find any USB ports on the system, it's best to stay away from USB. You can add a USB card, but getting it to work right can be tricky.

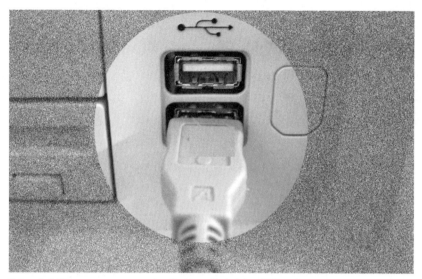

Figure 5-4 The newer your computer, the more likely it has USB ports.

If you have USB ports, but they're all taken up with things plugged into them, that's not a problem. One of the USB devices you already have may serve as a hub, with additional connectors for accepting other USB devices. USB keyboards, for example, often include USB connectors.

Check out your monitor also. Some monitors have built-in USB hubs, even though the monitor itself doesn't use USB for anything. If your monitor has USB connectors, it should have one for plugging in a cable between your computer and the monitor (an *upstream port*) and additional ports for plugging in other USB devices (*downstream ports*).

If you don't have any available USB ports, and none of the USB devices you already have also functions as a hub, you can buy a hub, like the one in Figure 5-5. Like the monitors with built-in hubs we just described, the hub will include an upstream port for connecting to your computer and provide some number of downstream ports for connecting other devices. The version in the figure, for example, offers four downstream ports. Simply disconnect one of the devices you have plugged into a computer USB port, plug in a cable between the computer and the hub, and then plug the cable you just removed from the computer into one of the USB hub connectors.

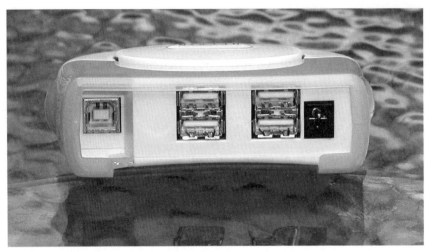

Figure 5-5 USB hubs let you add more USB ports.

By the way, note that USB hubs come in both powered and unpowered versions. (The powered versions include a power cord to plug into an outlet.) A powered version is generally the better choice, since some USB devices won't work if you plug them into an unpowered hub.

An alternative to a card reader is an adapter that will fit in a slot you already have in your computer. The two slots we have in mind are your floppy disk slot, which you should have available in virtually any computer, and the PC card slot, which you probably don't have on your desktop system, but almost certainly have on your notebook, if you have one. Figure 5-6 shows both kinds of adapters.

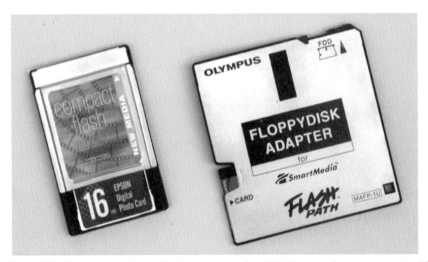

Figure 5-6   You can find adapters like the one on the left, which lets you read memory cards in a PC card slot, or the one on the right, which lets you read memory cards in a floppy disk drive.

Floppy disk adapters, which we've seen in versions for SmartMedia cards and Memory Sticks, let you put the memory in the adapter, put the adapter in your floppy disk drive, and read from the card as if it were a floppy disk. However, as of this writing, they tend to be more expensive than memory card readers, so we'd recommend getting a card reader instead. If you get a floppy disk adapter, make sure it will work with your version of Windows.

We've seen PC card adapters for CompactFlash, SmartMedia, MultiMediaCard, Memory Stick, and Secure Digital cards, as well as for the IBM Microdrive (a 1-inch hard disk in a CompactFlash-size card). We've even seen a single card that can handle more than one format of memory card. If you have a notebook system with a PC card slot, or a PC card adapter on your desktop, getting a PC card adapter for your camera memory can be less expensive than getting a card reader.

One other option we should mention here is getting a photo printer that can read memory cards. Some photo printers include slots for one or more types of cards. In some, but not all, cases the printer is set up so the computer it's attached to will see the memory card as a drive, essentially letting the printer double as a memory card reader. We'll cover this in more detail a little later in this chapter. For the moment, just keep it in mind as one more alternative for physically connecting between your memory card and your computer. If you're thinking of getting a photo printer, you may wind up with what amounts to a memory card reader as a bonus.

More Choices: Moving Photos Versus Moving Files

Getting a physical connection between your memory card and your computer is only half the battle. You also have to move the files from the card to your computer.

Most, if not all, cameras come with some sort of software for transferring files from the camera to your hard disk. Typically, once you install the software it will recognize the camera when you plug it in, let you browse the pictures that are in the camera memory, and let you copy the pictures over to the computer, where the program can store them in any number of ways. The manual that came with the camera should explain the details.

Whatever the details of moving the photos to your computer this way, you're basically treating the photos as photos rather than files. The program may well store each photo as a separate file, but it could just as easily store it as part of a larger file that contains all your photos, and you wouldn't know it. If the program lets you put photos into individual albums, or provides some similar way to categorize the photos, it may create a separate folder on your disk for each album you create in the program, or it may throw all the files into a single folder on disk and keep track of which photo goes in which album strictly by keeping track of file names. If that's the strategy it uses, the file names may or may not match the names you use for the photos when you're working within the program.

The problem with relying on a program like this to move and track your photos is that you don't really know what's going on with your photos. It's a little like having a car with the hood welded shut. When the time comes to check the oil—or, in this case, to find a photo you want to edit or insert into a word processing document—you not only have to crack the hood and go into unfamiliar territory, but you may have trouble finding what you're looking for.

The alternative that many camera manuals don't mention is that you can simply treat the photos as files. That's the way they're stored in your camera's memory, with each photo saved as a separate file. You can use Windows Explorer or any other program you're comfortable with to copy them from the camera's memory card to anywhere you care to keep them on your hard disk. You can rename the files, create folders on your disk to organize the photos into groups, and otherwise manipulate the individual files or folders. If you're using Windows XP or Windows Me, you can even see thumbnails of the files in each folder, as in Figure 5-7. Simply go to the folder the files are in, open the View menu, and choose Thumbnails.

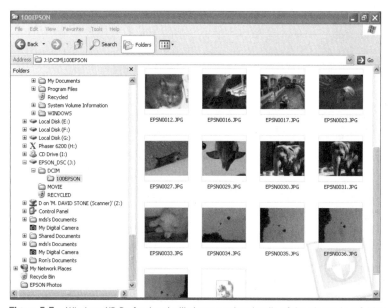

Figure 5-7   Windows XP Professional will show you thumbnails of the photos in a folder.

Most current camera models will let you treat the memory card as a removable disk simply by connecting a USB cable between the camera and the computer. Windows XP and Windows Me will recognize the camera memory as a USB mass storage device and add it as a drive that you can get to from My Computer, Windows Explorer, or any other program. Figure 5-7, for example, shows an Epson camera (EPSON_DSC) as drive J. The thumbnails on the right side of the window are actually files that are on the camera's memory card, which is sitting in the camera. You can simply go to the memory card in Windows Explorer, navigate to the folder with the files, and look at them as thumbnails, copy them, or move them to your hard disk.

Windows XP and Windows Me include drivers on their distribution discs for treating some cameras as drives. In those cases, when you plug the camera in the first time, Windows will automatically install the appropriate driver. With other cameras and with other versions of Windows, you'll need to install a driver that comes with the camera.

Alas, some cameras aren't designed to work as USB drives. Older models that use the standard serial port can't pull off that trick. Some later models that use the USB port don't try. In those cases you can still get Windows to recognize the memory card as a drive by moving it to a card reader or to an adapter. If your version of Windows works with the card reader or adapter, Windows will treat the card as a drive. Once you get that far, you can copy and move the files at will.

Storing Your Photos

Even low-resolution compressed photos take up room, which means that once you get enough photo files on your hard disk, you'll inevitably start wondering how to make enough room for them. And the more photos you take, the sooner you'll need to wonder about that.

Basically, you have two choices for storing photos: you can keep getting bigger hard disks, or you can move the files to removable *disks* or *discs*. (The first spelling indicates a magnetic disk, like a Zip disk. The second indicates an optical disc, like a CD or DVD.) Or you could do both.

Lingo   *Disk* is the right spelling for magnetic recording media like hard disks, floppy disks, or Zip disks. But *disc*, ending with a *c*, is the right spelling for optical discs, like CD or DVD discs.

The Hard Disk Option

Hard disks have been growing in capacity and dropping in price per megabyte so quickly for so long that it's not even news any more. We can safely say that, well into the foreseeable future, whatever size hard disk you have on any given date, if it's more than a year old, you'll be able to get one with much more capacity for a price that will surprise you.

You don't even have to know much to install a hard disk. If you don't want to open your computer's case to add to or replace your current drive, you can get one that plugs into a USB port or into a FireWire port. (FireWire, or IEEE 1394, is similar in many ways to USB, but isn't as widely used, so your computer is less likely to have it. There's no need to go into details here; just make sure you match any external hard disk you get to the type of port you have available on your computer. Everything else is pretty much automatic, and you should get any installation instructions you need with the drive.) External drives that use a USB or FireWire port will be slow compared to internal drives, but they're plenty fast enough for storing and retrieving photos.

The advantage to keeping all your photos on a hard disk is straightforward. You'll have all your photos available all the time, without having to hunt for a disk or disc when you want to retrieve a photo, not to mention the chore of finding the *right* disk or disc. The disadvantages are pretty straightforward too. You'll be using up room on your hard disk, and if the hard disk dies, you'll lose all your photos.

Using up room on your hard disk is not as much of a problem as it once was. You have to take a lot of pictures before it becomes an issue.

A couple of examples will help make the point. Let's start with an unlikely scenario that will give you gigantic files compared to what you'll have in real-world use. Suppose you have a 3-megapixel camera and choose to use no compression and the highest resolution at all times. Each photo will be a bit more than 9 MB, giving you room for about 11 photos per 100 MB, or 110 photos per gigabyte. That means a 20-GB hard disk, which isn't terribly large by today's standards, can hold 2200 photos.

If you use the highest resolution for the camera, but turn on compression, each photo will vary in size, depending on the amount of detail in the photo. But even with minimal compression, you'll likely need well under 1 MB per photo. To make the math easy, assume each photo is 1 MB. In that case, the same 20-GB hard disk will hold 20,000 photos. By the time you take that many, hard disks with much greater capacity will be available for far less money. Use a lower resolution or more compression, and it will take even longer to fill up the 20 GB.

The real problem with hard disks is that they die. It doesn't happen very often, but it does happen. Hard disks have gotten reliable enough so that too many people have forgotten they can die. Don't be one of them. The truth is that hard disks are just reliable enough to lull you into a false sense of security.

When a two-year-old, 60-GB hard disk simply stops working without warning (as happened to one of us recently), the drive will probably still be under warranty, and getting a replacement will be easy, but everything on the drive will be gone. You don't want that to include 20,000 photos—or even 100 photos that you care about—that you've gathered over a period of years. The way to avoid that problem is to keep your photos in a more reliable storage format.

The Removable Disk (or Disc) Option

An alternative to storing your photos exclusively on a hard disk is to keep them on some sort of removable disc (or disk), either instead of keeping them on the hard disk, or in addition to it, in which case the copies on the removable discs or disks are your backups.

We're talking primarily about (optical) discs here, rather than (magnetic) disks, because optical discs—which today means one of the writable versions of CD or DVD—are the preferred choice. Odds are you already have a *CD-R/RW drive*, often called a *burner*, in your computer for writing to CD-R and CD-RW discs. If you don't, you should seriously consider getting one—or at least plan to get one on your next computer. They're not expensive, and they are extremely useful, not just for photos, but for moving large files from one computer to another. There are several varieties of writable DVD drives, but at this writing they're a bit expensive.

Lingo   A *CD-R/RW drive*, often called a *burner*, is a drive that can write to CDs, using either CD-R or CD-RW discs. Unlike CD-R discs, CR-RW discs can be erased and rewritten.

The reason we prefer optical discs to magnetic disks is, quite simply, they keep your data intact longer. Unlike magnetic disks, they aren't subject to the vagaries of a stray cosmic ray (we're not kidding) flipping a bit and making a file unreadable, as happens with magnetic disks all the time.

You want to be able to put your photos on a disc, put the disc on a shelf, and know that the file will still be there when your grandchildren's grandchildren rummage through your old photos decades from now. With an optical disc, you can be reasonably sure the data will still be there. With a magnetic disk, you can be reasonably sure that it won't be.

Caution   When it comes to longevity, all optical discs are not created equal. Longevity for CD-R discs varies from about 60 years to over 200 years. The verdict is still out on CD-RWs. When you buy discs, check the manufacturer's claimed longevity.

Our advice: if you're keeping your photos on your hard disk and using removable disks strictly for backups, there's no problem with using Zip disks, Jaz disks, or other magnetic disks. But if you're storing your photos long term, or not keeping copies on your hard disk drive, optical discs are the way to go. Make that: the only way to go. (And if you're thinking about using tape, note that we left tape out on purpose. It has all the drawbacks of magnetic disks and isn't as convenient for retrieving files.)

The Third Way

There's one other choice for storing your photos that we have to mention. There are some Web sites, notably *www.photoaccess.com*, that let you store your photos online. At this writing, the storage is unlimited and free (something that may change if everyone tries storing thousands of photos). The hope, presumably, is that by providing the free storage service, the site will draw people in for paid services.

We have some reservations about storing your photos this way, but we'll discuss them in more detail and take a closer look at the sites themselves in Chapter 13, "Sharing Your Photos: E-mail, Letters, and Web Sites." For now, it's enough to know that storing your photos online is an option.

There's Something about Printers

In addition to taking and storing pictures, you'll almost certainly want to print your pictures. And that means you have to know at least a little about printers. We aren't going to try to tell you everything there is to know about printing— that's enough material for a book of its own. Our intention here is to sketch out the highlights so you can understand the limitations you may run into with the printer you already own and make better choices about your next printer when you decide it's time to get one.

What Makes a Printer a Photo Printer

Most inkjets that are less than a year or two old can do a reasonably good job of printing photos. Some can do an outstanding job. The best will print photos that we defy anyone to tell came from a digital printer instead of a photographic print. Even many color laser printers today can do a reasonably good job with photos. All of which may make you wonder if there's any point in getting a photo printer. Or even more pointedly, you may wonder what makes a printer a photo printer.

Ultimately, a photo printer is anything that a manufacturer decides to call a photo printer. Most often, however, the term means that the printer has some special feature that gives it an advantage for photos, either because of the printing technology, the logistics of handling photos, or both. Here's a quick rundown on the three most common reasons a manufacturer will add *photo* to the model name.

Thermal Dye Printers

Thermal dye printers are probably better known as dye sublimation printers, although that's technically the wrong name for them. By whatever name, however, they are known for printing photos well, and they aren't good for much else. They print on a coated paper stock that looks and feels much like photographic prints, which limits their usefulness for other purposes. They work by heating a ribbon so that dye in the ribbon will migrate into the coating on the paper. Their key strength is that they can print each individual dot on the page in any of 16.7 million colors, which lets them print continuous tone images. The images they print also tend to last longer without fading than images printed on inkjets.

The thermal dye photo printers of most interest to nonprofessionals are relatively small and are limited to small output size, often as small as 3.4 × 2.1 inches (wallet size), 3 × 4 inches, or 4 × 6 inches. If you're serious enough about photography to invest in a printer that does nothing but photographs, you might want to investigate thermal dye printers. Similarly, if you absolutely have to have a portable printer for high-quality photo output, a thermal dye printer may be your best choice. But for most people, there's little point in bothering with anything but inkjets.

Six-Color Inkjets

Most inkjets—indeed, most color printers, including thermal dye photo printers—use just four color inks: cyan, yellow, magenta, and black. As it happens, however, with inkjet technology, you can get finer control and more natural-looking colors in things like skin tones and the shades of blue in the sky if you add two extra colors: light cyan and light magenta.

Not surprisingly then, most photo inkjets are six-color printers. If you're shopping for a new printer, however, keep in mind that having six ink colors instead of four doesn't guarantee better looking output; it just increases the odds. What matters is what the result looks like, not how the printer got there.

No Computer Needed

Some photo printers give you a way to print photos without going through a computer. In some cases, the printers are designed to work only with specific cameras (from the same manufacturer, naturally). You plug in a cable between the camera and printer, and then give a command to print. More often, however, the printer includes a memory card slot, so you can take the memory card from any camera that uses the right type of card, put it in the printer's slot, and print.

Figure 5-8 shows the Epson Stylus Photo 785EPX, complete with its liquid crystal display (LCD) preview monitor—the small box sticking out on the top right—for previewing photos before you print them.

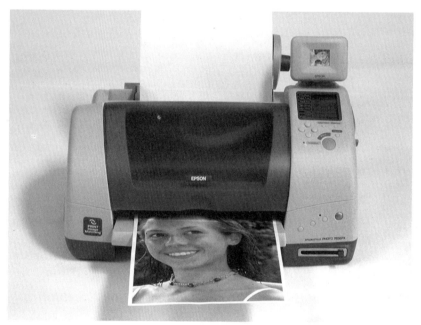

Figure 5-8   The Epson Stylus Photo 785EPX earns the term photo printer, in part, by letting you print without a computer.

The printer is as much a miniature photo lab as it is a printer. The slot at the bottom right takes a PC card, which means the printer can accept any kind of digital film that there's a PC card adapter for—CompactFlash, SmartMedia, Memory Stick, MultiMediaCard, Secure Digital, or IBM Microdrive. To print from a card, you first insert the adapter into the slot, then use the control panel just below the preview monitor to set print mode, paper size, quality, which files to print, and so on. Then you give the command to print.

Be aware that some printers that include memory card slots will let you use the memory card only to print. Others, including the Epson 785EPX, will let your computer use the slot as a memory card reader so you can move the photos to your computer. (And yes, in case you're wondering, you can plug the 785EPX, as well as most printers with memory card slots, into your computer to use them as standard printers, too.)

About General-Purpose Printers

Before you get too carried away with the idea of photo printers, keep in mind that most inkjet printers today can produce reasonably good looking photos. If you've tried printing on your inkjet printer and you're not satisfied with the quality of the photos you're getting, don't give up too soon.

Tip   The most recent color lasers are getting pretty good at photos, too—something worth knowing if you have a color laser printer available.

Whether you're using a standard inkjet printer or a photo printer, you won't get the best results unless you use the right paper, set the printer for the paper you're using, and check the printer driver for options that can improve your output (something that most people never bother to do). You may even need to switch to a different set of inks. Some printers have special photo ink, as distinct from the ink they use for standard printing. We'll cover all of this in Chapter 11.

Printer Limitations

There are a couple of issues related to printers that you should be aware of. At best, if you understand why you're having a problem getting reasonable-looking colors in your printed output, or why the printed photos fade, you may be able to do something about it. At worst, knowing why you're having the problem may keep you from spending a lot of time wondering if there's something wrong with your printer or, worse, pulling out your hair trying to fix a problem that can't be fixed.

When Colors Don't Match

In Chapter 1, in the section "Resolution and Film," we pointed out that getting a reasonable match between the color the camera sees, the color that shows on a monitor, and the color that prints can sometimes be a problem. As we said in that discussion, color matching is an issue for everything from film prints to matching colors in rolls of wallpaper. The problem for digital photography is that when you print the photos yourself, you're the one in charge of getting the colors right.

There are all sorts of reasons why getting colors to match isn't the simple task most of us think it should be. One of the most important issues is that printers can't print all the colors you can see in the real world, or even all the colors you can see on a monitor. (And monitors can't show all the colors a printer can print for that matter.)

The second big obstacle to getting colors to match is that printers, cameras, and monitors each have their own internal language (for lack of a better term) for defining colors. Your printer will define a given color as some mix of cyan, yellow, and magenta. A different printer, using different pigments or dyes, will define the same color as a different mix of cyan, yellow, and magenta. Your camera, meanwhile, will define the same color as some mix of red, green, and blue.

When your computer translates from the camera's language to the printer's language, it may or may not make allowances for the differences between one printer's definition for each color and another printer's definition. If it has what amounts to a translation table—a Rosetta stone for the specific camera and printer models—it can translate to match the printer's definition and the printed colors will come close to the original. If the computer doesn't have that translation table, the colors can be way off.

Color matching technologies are all about getting the translation right. The techniques themselves aren't all that difficult to add to printers, cameras, and software, but until recently each manufacturer went its own way on color matching. Now, all of them have largely agreed on which approach to use, and it's getting easier to pick a camera and printer at random and still get reasonable colors in your printed photos. If you have an older camera or printer, however, it may not share the newest technologies that help make the colors close enough for most purposes.

If you run into that problem, you may want to experiment with a newer model camera, printer, or both to see if getting new hardware will solve the problem. If it does, it's time to think about upgrading. Whether it solves the problem or not, you'll certainly want to get familiar with any features in your graphics editing program that will let you adjust the color in your photos. We'll discuss some of these possibilities in Chapter 9, "Advanced Editing: Fixing Flawed Photos" in the section "Adjusting Color."

Diamonds Are Forever, But Printed Color Changes

As you may have noticed if you have some old snapshots sitting around, color in photographic prints shifts hue and fades over time. In fact, if you've ever tried to patch a flaw in paint that has been on your wall for a few years with the same paint that you've stored in a can, or you've moved a piece of furniture that was covering up a part of your carpet, you've probably noticed the same thing. Colors change when they're exposed to light and air.

The same thing happens with printed photos, but it often happens much faster. We've seen inkjet-printed photos fade beyond the point where you'd want them hanging on your wall in just two or three years. And some combinations of ink and paper will shift color noticeably within a matter of weeks if they're exposed to certain pollutants. You could use them for testing air quality.

Printer manufacturers are continually reformulating their inks and papers to produce colors that not only look better, but last longer. Obviously, you don't have any control over that. But there are some things you can do to make your printed photos last.

First, treat third-party inks with suspicion. Printer manufacturers develop their ink and paper in combination with each other, and it's the combination of the two that produces both the color you see and the longevity of the prints. We don't pretend to know all the details, but we know enough to know that the physical (as in physics) and chemical properties of the ink and paper interact with each other to make a difference in what you see. That's why you'll see significant differences in printed output using the same printer and ink on different paper stock. Even glossy photo paper and matte photo paper from the same manufacturer will produce subtle differences in colors, and sometimes obvious differences in longevity.

Replacement cartridges or cartridge filler kits from other manufacturers have to be reverse-engineered for the printer. There's a lot of work involved in getting it right. And there's no guarantee that it will last as long as the printer manufacturer's ink. (Of course, there's always the possibility that it will last even longer, but given that the printer manufacturer is responsible for the paper too, you know who has the inside track on getting the ink right.)

Second, check to see if the manufacturer offers any special instructions about preserving prints from your printer. It may suggest that you use a specific paper or avoid using another paper. Or it may advise against using a particular combination of ink and paper unless you frame the photo, with the photo under glass to protect it. The most likely places to find suggestions like these are in the printer manual or on the manufacturer's Web site. But don't overlook the possibility that they're printed on the package the ink or paper comes in, or in an insert in those packages.

If the manufacturer doesn't offer any suggestions, you might want to experiment with different papers to see if some provide longer lasting prints than others. (The ink and paper combination we mentioned earlier, which can shift color in a matter of weeks, happened only with that specific combination of ink and paper.) You can also try searching the Web for this information, using your favorite search engine (*www.altavista.com, www.google.com,* or *www.yahoo.com,* for example). There are any number of professional photographers who care very much about longevity for their prints, and share what they've found on Web sites. Try searching for the printer model along with some combination of words like *paper, longevity, print life,* and *color shift.*

Finally, make sure the original digital files are stored safely. One of the more important benefits of digital photography is that you can always print out a new original, and it will be brand new and clean. Film negatives fade also. Digital files don't.

Key Points

■ You can physically connect a memory card to a computer for transferring files in a number of different ways. These include connecting a cable between the camera and computer, adding a memory card reader to your computer so you can move the card to the reader, and moving the memory card to an adapter that fits into a floppy disk drive or into a PC card slot.

■ Most cameras come with software for transferring photos to your computer, but it may be easier to use Windows Explorer or another program to copy or move them as files from your camera to your hard disk.

■ Modern hard disks offer plenty of room for storing lots of photos, but if the hard disk dies, you could lose all your photos. It's best to store the files on removable discs or disks, either instead of or in addition to storing them on your hard disk.

■ Optical discs, like CD-Rs, store files more reliably for longer periods of time than magnetic disks, like Zip or Jaz disks.

■ Matching the colors you print to the colors the camera sees is not as easy a technological trick as most people assume. The newer the camera and printer, the more likely they are to follow the same standards to match colors.

■ If the colors don't look right when you print them, you may be able to fix the color in a graphics editing program.

■ Colors in photographic prints fade. So do colors in inkjet prints. Check the printer manufacturer's site to see if it has any hints about how to increase the life of your prints, like using a specific combination of paper and ink.

Chapter 6

Keep Those Pictures Coming: Batteries and Digital Film

Taking pictures with abandon is easier said than done. The problem is simple: batteries run down and digital film fills up. Battery technology is the bane of digital cameras. And the batteries always seem to run out just as you're framing a great shot.

Even when your battery isn't running out on you, your digital film might. Most cameras come with minimal memory. If you use a high-resolution mode, you'll find the memory can fill up with as little as one shot. Fortunately, there are some things you can do to make batteries last longer and there are ways to fit more images on one piece of digital film. We'll tackle batteries first.

Batteries Included

Odds are that you don't pay much attention to batteries. You've used them for as long as you can remember, you're thoroughly familiar with them, and you don't think there's much of anything to pay attention to. We're going to try to change that attitude.

Your camera probably came with batteries. If they were standard AA batteries, you've probably long since used them up and thrown them away. If they were rechargeables, they're probably still going strong. And therein lies the first lesson you need to learn about batteries: not all batteries are created equal.

Batteries can be unequal in all sorts of ways. Whether you're talking about rechargeable or nonrechargeable batteries, some technologies actually do last longer than others or have other advantages. Even something as basic as nonrechargeable AA batteries come in both garden-variety versions and special versions that can power a digital camera for noticeably longer. Figure 6-1 shows some of the choices.

Figure 6-1   These are all AA batteries, but one is a high-powered alkaline, one is a standard alkaline, one is a NiCad, and one is a lithium battery.

Table 6-1 shows the most common choices in types of batteries and two of their key features: the number of recharge cycles (how often you can recharge them) and whether they suffer from a *memory effect*, which is basically a use it or lose it rule for battery life. If you recharge the battery when it still has some level of charge, the battery may treat that level as the point to go dead next time.

Effectively, it remembers the level it was at when you recharged. If you use a type of battery that suffers from a memory effect, you have to fully discharge the battery before recharging at least on a regular basis, if not every time.

Lingo   Some battery technologies have a *memory effect*, so if you don't discharge the battery fully, it will remember the charge level that you started recharging from and stop supplying power when you reach that point again.

Table 6-1   The Most Common Choices in Batteries

Battery Chemistry	Recharge Cycles	Memory effect
Alkaline	0 (not rechargeable)	N/A
Lithium	0 (not rechargeable)	N/A
Nickel cadmium (NiCad)	500-1000 recharges	Yes
Nickel metal hydride (NiMH)	500 recharges	Yes (but not as much as NiCad)
Lithium ion (Lion, or Li-ion)	500-1000 recharges	No
Lithium polymer	500-1000 recharges	No

In addition to varying in the number of recharge cycles and whether they have a memory effect, battery technologies differ in how quickly they discharge if you leave them sitting around and in how much energy they can hold per unit volume and per unit weight.

Your ideal battery will allow plenty of recharges, have no memory effect, pack a lot of energy per unit weight and per unit volume, and will not discharge significantly if you leave a charged battery sitting on a shelf.

In reality, you're limited to the batteries actually available for your camera. Keeping that in mind, here's a look at the six types of batteries we listed in the table. We start with the least desirable and work our way to the most desirable. If your camera allows more than one choice, pick one as far down the list as possible. (You may have noticed that we did not include rechargeable alkalines in the table. That was not an oversight. Don't even consider rechargeable alkalines: they don't last as long as regular alkalines, they take a long time to recharge, and they allow relatively few recharges.)

Caution   Don't assume that just because a battery fits in your camera that it's okay to use it. If the manual doesn't say it's okay to use a particular type of battery, assume it's not until you find out otherwise.

Alkaline Batteries

Alkaline batteries are the basic, nonrechargeable choice that you can find in any supermarket, grocery, or stationery store. They're cheap and widely available, but they're not a great choice for digital cameras. Our experience is that they don't last long enough with digital cameras to be truly useful.

We can remember testing digital cameras not all that many years ago with standard alkaline batteries, and finding that the batteries would die after just one or two shots. Literally. One or two shots. Granted, the shots were indoors, which meant they used flash, and we also had the liquid crystal display (LCD) on. And we hasten to add that it didn't happen with all cameras. But it wouldn't have made a very persuasive battery commercial.

Tip   Because cameras are so power hungry, batteries can fall below the level a camera needs long before the batteries are actually dead. You can often continue using them in other devices, like desktop phones.

Since then, battery manufacturers have come up with better alkaline batteries that are usually tagged as being designed for digital cameras and other electronic devices. Duracell, for example, sells both a standard alkaline battery and the Duracell Ultra. (You have to read the labeling, however. Some Duracell Ultra batteries are lithium batteries, which is a different technology altogether.)

If you must use an alkaline battery, we strongly recommend that you use one of these more muscular versions. Given the choice, however, we'd urge you to get rechargeable batteries instead. Not only do rechargeable batteries last longer between charges than alkaline batteries last before needing replacement, they wind up being less expensive in the long run.

If your camera uses AA batteries and it came with alkaline batteries, don't buy any more replacement alkalines. Go directly to the closest store that sells rechargeable batteries and get some, along with a charger. Resort to alkalines (or the lithium batteries we cover next) only in emergencies—like when your rechargeable batteries are all drained and you realize you left your recharger at home.

Lithium Batteries

Don't confuse lithium batteries with lithium ion or lithium polymer batteries. The second and third versions are rechargeable. The first is not. Ultimately, lithium batteries are in pretty much the same category as alkaline batteries, except that they cost more, last longer, and aren't as widely available.

If you think of lithium batteries as still more muscular variations on alkalines, you won't be far off. More important, our advice about alkalines applies to lithium batteries also. Rechargeables are a better choice. If you must use a non-rechargeable battery, choosing between lithium and alkaline is a purely economic decision that may come out differently depending on how many more shots you plan to take before you can get your rechargeables back online.

Our best advice is to get some experience with both lithium and alkaline batteries in your camera (assuming your camera can use both) to get a feel for how much longer lithium batteries last. You might even want to keep track of how many shots you take each time you use one type of battery or the other. Once you see the difference in battery life in your particular camera, you can decide whether the extra cost of lithium batteries is worth it on any given occasion.

Nickel Cadmium (NiCad) Batteries

The venerable nickel cadmium (NiCad) rechargeable battery will probably be around for a long time, since it's less expensive than the better choices, like nickel metal hydride (NiMH). NiCads are best known for their disadvantages, including their memory effect and relatively short battery life compared to other rechargeable options.

As we mentioned earlier, the memory effect means that if you don't fully discharge the battery before recharging, you may find that the battery remembers the point where you started recharging. If that happens, it will treat that point as the level to stop supplying power. The higher the level that it treats as zero, the less time it will give you before you have to recharge again. And if you keep recharging too soon, you may find that you have little or no running time before the battery stops working.

There is some disagreement about how significant the memory effect is for current incarnations of NiCads. Whatever the truth is for current designs, it's certainly true that NiCad batteries are much more prone to the memory effect than other kinds of batteries. But they have their good points too. NiCads hold their charge reasonably well if you leave them sitting around, with a self-discharge rate of about 15 percent per month.

Nickel Metal Hydride (NiMH) Batteries

Nickel Metal-Hydride (NiMH) batteries offer more energy per ounce and per unit volume than NiCads (read: they run about 30 percent longer). They are also less susceptible to a memory effect. However, they have a tendency to self-discharge relatively quickly—at about 30 percent per month—if you leave a charged battery sitting around, and you can recharge them only about 500 times, compared to as many as 1000 times for NiCads.

All this makes NiMH batteries preferable to NiCads overall, but less than ideal. Even so, a fair number of people consider NiMH the best current compromise for digital cameras between battery features and price. Don't count us in that group, however. Given the choice, we'd rather have a lithium ion or lithium polymer battery. (Note, however, that you make this choice when you buy the camera. Any given camera will come either with a lithium ion or lithium polymer battery on the one hand, or will let you use batteries with one or more of the other battery chemistries. At least, we've never seen a camera that lets you switch back and forth.)

Lithium Ion Batteries

All other things being equal, we'd take a camera that uses lithium ion batteries over one that's limited to, say, alkaline, lithium, NiCad, and NiMH. Lithium ion batteries cost more than even NiMH batteries, but you can recharge them more than 1000 times compared to 500 times for NiMH; they pack about 80 percent more energy than NiMH batteries per unit weight; and if you store them they lose power at a rate of only about 8 percent per month. They also eliminate the memory effect as something to worry about.

Lithium Polymer Batteries

Lithium polymer batteries are very much in the same category as lithium ion batteries. They offer essentially the same advantages, but they're made of a material that manufacturers can shape to fit all sorts of odd spaces. That makes them attractive as proprietary batteries for digital cameras, since designers may need to shove the battery into whatever space is available.

Maximize Your Battery Life

Whatever kind of battery you wind up with, you'll want to keep it running as long as you can. There are actually two categories of things you can do to maximize battery life. First, most people don't realize it, but there are ways to take care of your batteries so they last longer. Second, there are features you can turn off or use sparingly to minimize the drain on your batteries, so they'll last longer. We'll look at each category separately.

Treat Your Batteries Well

When you buy rechargeable batteries, or get some with your camera, make sure you read the instructions that came with them. The most critical issue is whether there is some conditioning you should do when you first get them. For example, we've seen instructions to charge your batteries before you use them the first time, give them a chance to cool down, then charge them again, give them a chance to cool down again, and then charge them again. More common is the instruction to charge them and then run them completely into the ground before recharging at least the first two times, and in some cases the first three times, that you charge them.

Whatever the directions for conditioning, if you follow them, the batteries will perform better and live longer. If you didn't get any directions with the batteries, but they have a recognizable manufacturer name and model, you can try looking for the battery model on the manufacturer's Web site, and see if there are conditioning instructions there.

There are also things you can do after the initial conditioning to maximize battery life. For NiCad and NiMH batteries, it's a good idea to run them down completely at least once a month to avoid problems with the memory effect. Your battery recharger may include a run-down option. If not, you can find products that will fully discharge the battery for you. It's also a good idea to avoid recharging until you need to. When you recharge, you're using up a recharge cycle, and coming one cycle closer to not being able to recharge anymore.

Lithium ion and lithium polymer batteries need very different treatment, and often come with instructions to charge the battery every chance you get. With no memory effect, there's no need to run the batteries down, ever. Keep in mind, however, that the best way to maximize battery life varies even from one battery model to another. The rule remains: read the directions that came with the batteries.

There are also some dos and don'ts to keep in mind about batteries:

- Do make a point of replacing all your batteries at the same time. Most people do this pretty much automatically with rechargeables, but even with nonrechargeables, swap out the entire set of batteries in the camera at once. If you leave one partially drained battery in the camera and put three new batteries in, the three new batteries will drain much more quickly than if you have four matched batteries to start with.

- Do make a point of using four identical, new batteries when you replace them. Don't mix batteries from different manufacturers or different models from the same manufacturer. Mixing batteries can create conditions that cause the batteries to leak, which can damage your camera.

- Do make a point of storing batteries in a cool, dry place at normal room temperature.

- Do make sure that the contacts on your rechargeable batteries and in your camera stay clean and make good electrical contact. If they seem dirty, use a pencil eraser to clean them.

- Do make a point of giving rechargeable batteries a chance to cool down after recharging instead of popping them into your camera immediately.

- Don't leave batteries—or your camera with batteries in it—sitting in places like a hot car in direct sunlight for extended periods of time. Heat raises the self-discharge rate. (And it isn't a good thing for the electronics inside your camera, either.)

- Don't leave batteries—or your camera with batteries in it—sitting in places like a cold car sitting in an unheated garage in the middle of a New England winter with temperatures at 17 degrees below zero. Cold also drains batteries quickly.

- Don't try to recharge batteries in a charger that's not designed for them. Match the charger to the battery type.

- Don't be cheap. Buy an extra set or two—or three—of rechargeable batteries, and charge them all to bring with you. When one set dies, start recharging it if you have a place to plug in, and move on to the next set. By the time you've run through all your batteries, the first set should be recharged.

Minimize the Drain on Your Batteries

After you put your batteries in the camera, there are still some things you can do to maximize battery life. At the top of this list, and deserving special mention, is not to use your LCD any more than you have to.

As we've pointed out elsewhere in this book, most notably in Chapter 2, in the section "What's SLR, and Why Does It Matter (But Maybe Not as Much as

You Think)?" framing your pictures by using the LCD instead of the viewfinder will help you avoid doing things like chopping off the top of people's heads in your photos. However, as we've also pointed out, using the LCD will shorten your battery life.

The trick here is to use the LCD only when you need to, and turn it off otherwise. Don't use it for shots where it isn't critical to frame the image just so; don't use it in sunlight that's so bright that you can't really see the scene on the LCD in any case; don't spend a lot of time admiring your handiwork in playback mode; and if you have a choice between changing a setting using the LCD menus or using a button on the camera, learn how to use the button so you can keep the LCD off.

More generally, any feature that requires power is a power drain, and the more of those features you can do without, or turn down to use less power, the longer your batteries will last. Here are some specific things you may be able to do, depending on your camera:

- If your camera has a sound recording feature, turn it off if you're not using it, or at least turn it to as short a time as the controls allow.

- For those times when you must use the LCD, check whether you can set the brightness or contrast on the LCD. (On an LCD, these two controls are actually two names for the same thing.) If you can set it, turn it down as far as you can and still see the display. A lower setting uses less power.

- If your camera has an automatic power-save mode, keep the timer at the lowest setting so it will power down sooner rather than later.

- If your camera lets you control the sounds it makes, turn off any sounds that you don't absolutely need, and turn the rest to minimum volume.

- If your camera has a jack for a power cord, as shown in Figure 6-2, get an appropriate power block if the camera didn't come with one, and use it instead of the batteries whenever possible. When you're scrolling through the photos stored in your camera, for example, trying to decide if you can delete any to make more room, you may be near a power outlet that you can plug into conveniently. Similarly, when you plug in a cable to move files over to your computer, you may find it convenient to plug in the power cord as well.

Figure 6-2   Most, but not all, cameras include a power connector for connecting to a wall outlet.

■   If you find yourself spending a lot of time looking at photos on the LCD just so you can decide which ones to delete to make more room, consider getting a storage card with more memory. If your camera came with, say, a 16-MB memory card, and you replace it with a 256-MB card, you may not have to worry about running out of room any more. And you can keep the 16-MB card handy as a spare, should you need it.

Making the Most of Your Digital Film

Some of the same tips that will lengthen battery life will also help you make the most of your digital film. The most obvious tip in that category is the suggestion to get a storage card with more memory, so let's start with that.

Digital film comes in more shapes and sizes than you may realize. Figure 6-3 shows some of the options.

Figure 6-3   These assorted memory cards, mostly from SanDisk, differ in size, shape, and capacity.

There are four kinds of cards in Figure 6-3. (And still other kinds that aren't included here. Some cameras even use floppy disks or writable CDs for their digital film, but those are the exception. The memory limitations we are about to discuss—and the tricks for getting around them—don't apply to those cameras because you can just replace the disk or disc.) If you look at them as loosely distributed around a clock face, the cards at 1 to 3 o'clock are MultiMediaCards, the one spanning 3 to 5 o'clock is a SmartMedia card, the one at 7 to 8 o'clock is a Memory Stick, and the two at 9 to 12 o'clock are CompactFlash cards. The more interesting information in the photo is the high capacity on some of the cards: one of the MultiMediaCards offers 256 MB of memory; one of the CompactFlash cards offers 512 MB. You can't read the capacities for the other cards in the photo, but the capacities are 128 MB for the Memory Stick and for the SmartMedia card.

Note   MultiMediaCard is often shortened to MMC, and CompactFlash is often shortened to CF.

We'll bet a chocolate ice cream soda that the memory you got with your camera is a lot less than any of the cards we just pointed out (unless this book is very old by the time you read this—and maybe not even then). As we write this, the typical digital camera comes with an 8- or 16-MB card. And that leads us to the first, and most obvious, recommendation for improving the number of shots you can get on one memory card: buy one with lots more memory.

How much more memory to get is partly a matter of what's available for the format your camera uses, but mostly a matter of budget. More is better, but a substantial amount of memory costs a substantial amount of money. That's why your camera came with so little memory in the first place, to help keep costs down. We recommend that you get as much as you can afford. Keep in mind that if your next camera uses the same kind of memory card, you'll be able to move the card to your next camera as well.

If your camera uses a CompactFlash card, don't overlook IBM's Micro-drive—a 1-inch disk drive mounted in a CompactFlash card format. As of this writing, the Microdrive offers more capacity than the highest capacity Compact-Flash memory card. Odds are that as time goes on it will stay ahead of the memory cards as well. However, before you invest in a Microdrive, make sure your camera will work with it. Not all do.

In addition to getting a larger capacity memory card, there are several things you can do to increase the usability of whatever memory you have.

- First, for any given picture, use the lowest resolution you need for however you plan to view the picture. We covered the requirements in detail in Chapter 3, in the section "Resolution and Compression Together."

  - Briefly, the rule of thumb for viewing on screen is to use one step lower in standard screen resolution than the resolution you expect to be using to view the photo. To view the photo on a computer using 1024 × 768 resolution, for example, you'll want the photo at roughly 800 × 600 pixels.

  - For printing, you'll want roughly 150 pixels per inch (ppi), or 200 ppi if you're more demanding. So a photo that you plan to print at 8 × 10 inches should have either 1500 or 2000 pixels in its longest direction.

- Use the highest compression level you're willing to accept for picture quality. We covered compression levels in Chapter 3, in the section "JPEG Format."

- Delete the photos you don't want. If you're on a dolphin watch and trying to take a picture of a dolphin jumping out of the water, delete the shots where all you get is a splash of water, and make more room on your card.

- Have a strategy for moving the pictures off the camera to give you more room. (We'll discuss some starting with the next section.)

Offloading Your Images on the Road

If you're away from home base, your camera's memory is filled up, and you need to free up some memory without deleting any photos, you have basically three choices: move the files over to a computer you own (such as a notebook you brought with you), move the files over to someone else's computer as a way station to another destination, or—the most interesting choice in many ways—find a kiosk that will let you burn a CD from your memory card.

Let's look at that last choice first.

Burn a CD at a Public Kiosk

Suppose you could walk into, say, a photo finishing store, a supermarket, or a drugstore, where you might drop off film for developing. But instead of dropping off film, you see a kiosk, like the one in Figure 6-4, half hidden in a recess and stuck between a watch display and a bunch of cardboard boxes with a stray bottle of glass cleaner. But it can print from digital film, so you head straight for the kiosk.

The particular kiosk in Figure 6-4 is from Kodak. As we write this, there are more than 21,000 of them scattered around the United States according to Kodak, and another 14,000 worldwide. Of the 21,000 in the U.S., roughly 80 percent can accept some type of digital film, primarily CompactFlash cards. By the time you read this, Kodak should be distributing new models that can read virtually every type of memory card available and let you print directly from your camera's memory card. However, printing the photos doesn't solve the problem of making room for more photos in your camera's memory. You'll want to keep those files whether you have a printed photo or not.

The part of the kiosk we're most interested in at the moment isn't available on the older models like the one in Figure 6-4, but it will be available on the new models that should be starting to appear around the time this book is published. It's a slot for burning a CD from your photos. Depending on when you read this, the kiosks equipped to burn a CD may still be rare. But they should be increasingly easy to find as time goes on, and you should be aware that they exist.

Figure 6-4   Kiosks like this can let you move your files to a CD.

In Search of a Kiosk

The Kodak Picture Maker kiosks are a great boon if you need to offload your pictures, but there's the little matter of finding one when you need it. That problem is partly solved by a store finder option that Kodak provides on its Web site. Keeping in mind that Web sites can change overnight, here's how it worked when we tried it.

Open your browser and enter the address *www.kodak.com/go/picturemaker*, which will take you to the Picture Maker home page. (We also got to the page by going to *www.kodak.com*, using the search function to search for Picture Maker, then picking the Picture Maker link from the list of hits. This might be handy to know if you forget the first address.)

On the right side of the screen is a list of options as shown in Figure 6-5. One of these is Store Locator. You can enter a city and state in the two text boxes, and then choose Go, but when we did that with several cities, we often struck out. (Well, okay; we made a point of picking places lost in the woods, but that might be just where you happen to be.)

Figure 6-5   You can use Kodak's Picture Maker Web site to find a nearby kiosk.

A more useful approach for us was to choose the Store Locator link just above the City and State text boxes. That takes you to a screen, shown in Figure 6-6, that will let you search by zip code. This obviously won't help much if you don't know the zip code for your current location. (If you're in a hotel or motel, try asking at the front desk.) But if you know the zip code, you can search wider and wider areas by using fewer and fewer numbers in the zip code.

Sometimes that's not necessary. Searching in 10016 (Midtown Manhattan in New York City) turned up the maximum of six locations the first time. But searching for 18944 (Perkasie, PA) turned up none. When we searched 1894, however, we got one hit, and when we searched for 189, we got the maximum six. If you try widening the search this way, make sure you go back to the screen in Figure 6-6 first. The next screen, which declares that no entries were found, offers a place to enter a new zip code, but it doesn't seem to change the zip code that the search is looking for. When we entered a new zip code on that screen, it still couldn't find any locations, no matter what we entered—including the zip code for Midtown Manhattan.

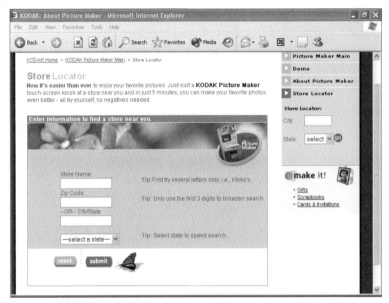

Figure 6-6   You can use this screen to search by zip code.

Here's another useful trick: once the search is broad enough to return the maximum six locations, repeat the search. We found that by going back to the screen in Figure 6-5 and searching again, we usually came up with six additional locations, giving us that much more to choose from.

Of course, if you can't get online, the Web site won't be much help. If you know that you might need a kiosk beforehand—if you're going on a vacation or business trip for example—get a list of locations from the Web site before you go. If you're already there and don't have a handy computer to get you online, you can call 800-939-1302 to get the information.

One last thing: when we went looking for a nearby kiosk, the first two stores we tried had nonworking units that were waiting for replacement parts. And the Web site, when we last saw it, did not identify which stores have digital capabilities. So before you go driving around, camera in hand, looking for a store, it's a good idea to call ahead and make sure the kiosk is working and that it can handle your type of memory card.

Taking Advantage of Your Computer (or Somebody Else's)

We covered the mechanics of moving files to a desktop or notebook computer in Chapter 5, "Special Issues for Digital Photography." For the moment, let's talk strictly about strategies.

If you brought a computer with you—a notebook for example—or you have an easy way to use someone else's computer, you can make room on your card by moving the photos to the computer. Given that moving files to a computer is what you would normally do with a digital camera in any case, this may seem too obvious to mention here, but we have some tricks up our sleeves. So read on.

Once you move the photos to a computer, you have several options for dealing with them. If it's your own computer, and you have enough spare room on the hard disk, you can leave the files sitting on the hard disk until you get back to your home or office, and can move them where you want to eventually keep them.

Note   You can find portable storage devices with names like the Digital Wallet and PicturePAD whose whole purpose in life is to provide multiple gigabytes of storage to offload your photos. However these aren't for casual users, since they cost more than many cameras.

If your computer is short on disk space, or you're using a computer that isn't yours, you can move the files to any removable storage that the computer offers—most often Zip disks or CD-R or CD-RW discs. (We covered these options in Chapter 5 also, in the section "Storing Your Photos.") Just make sure you're using a format that you can read on your computer at home base, whether home base is your home or office.

The next possibility is a little tricky and has limited usefulness, but it's worth keeping in mind if you have a low-capacity memory card and are working on someone else's computer: e-mail the photos to yourself. (We'll discuss the mechanics of e-mailing in Chapter 13, in the section "E-mailing Photos.") For details on the limitations and possibilities of e-mailing as a way to offload images while on the road, see the sidebar later in this chapter, "Waiting for E-mail."

Still another possibility that's somewhat limited, but worth considering, is moving the files to a personal digital assistant (PDA). Because a PDA is a lot more convenient to carry than a notebook, you're more likely to bring it with you. To make it worthwhile to move files to your PDA, you need one with at least 32 MB of memory. The harder part is that a good portion of the memory needs to be free. You also need a memory card slot in the PDA that matches the memory card your camera uses.

The memory requirement largely limits this trick to Pocket PCs, which tend to have more memory than other kinds of PDAs. Figure 6-7 shows a Compaq Pocket PC mounted in a sleeve that lets it accept a CompactFlash card. We'll discuss the mechanics of moving the files to the Pocket PC memory in Chapter 12, in the section "Putting Your Photos on Your PDA."

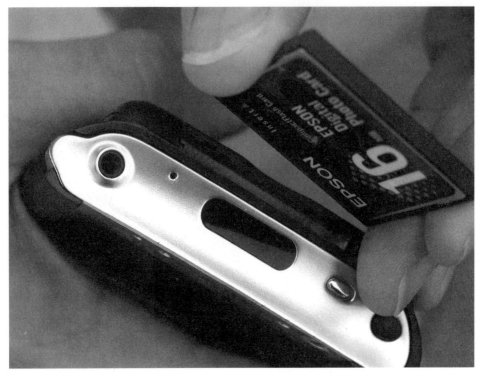

Figure 6-7   If your handheld has a slot for your memory card and enough free memory, you can move your photos to your PDA to make room on the card.

Waiting for E-mail The first limit on e-mailing to yourself is time and file size. Don't even think about e-mailing multiple megabytes worth of files unless you have a high-speed connection on both the computer you're sending from and the computer you'll be receiving the e-mail on. *High-speed* in this context means what's usually called a broadband connection, like a cable modem or DSL (Digital Subscriber Line). If you're using a standard modem and phone line to either e-mail the files or receive them later, it will simply take too long to make this approach reasonable.

Assuming you have broadband on both sides, e-mailing is still tricky, in part, because you want to send the photos without receiving them back on the system you're sending from. If you set up the system to send from your own e-mail account to the same account, you may wind up accidentally receiving the files to the system as well. And you could do that in a way that will prevent them from being available for you to receive again on another system, where you really want them.

The easy way to sidestep this problem is to send them from one account to a different account. You should have at least one account available from the company that you use to sign on to the Internet (that's your *Internet service provider*, or *ISP*). If your ISP limits you to one account, you can get another one for free from any number of Web sites, including Yahoo! Mail (available at *www.yahoo.com*).

Another issue that limits the possibilities for sending a large number of photos to yourself as e-mail is that most e-mail accounts have a maximum capacity for their inboxes. You can't send 600 MB worth of photos to an inbox with a maximum capacity of 5 MB. It helps to have at least a 25- to 50-MB capacity for your inbox. That's enough to be useful for cameras with 8- and 16-MB memory cards.

Fortunately, getting a reasonable storage capacity doesn't cost a lot. For example, Yahoo! Mail's free e-mail at this writing is limited to 4 MB of storage, but you can boost it to 25 MB for $19.99 per year (that's about $1.67 per month), to 50 MB for $29.99 per year ($2.50 per month), or to 100 MB for $49.99 per year ($4.17 per month). With 100 MB, you can clean out a 32-MB memory card three times before you run out of room in your inbox.

And in the word-to-the-wise department: don't count on sending lots of large files to your e-mail address at work. Your network administrator may or may not have a maximum limit defined for your mailbox. But unless he or she works for you, odds are you'll hear about it if you fill up the network server hard disk with your photos—even if they're business related.

Key Points

- Your camera may or may not have come with rechargeable batteries. If it didn't, get some. Use nonrechargeables only in emergencies.

- NiCads (nickel cadmium batteries) are the least desirable choice among rechargeables, largely because of the memory effect. However, they are also the least expensive choice.

- NiMH (nickel metal hydride) batteries offer better life between recharges than NiCads and are less susceptible to the memory effect, but they self-discharge relatively quickly, and can recharge only about 500 times.

- Lithium ion and lithium polymer batteries are the batteries of choice. They have little or no memory effect, pack much more energy than NiMH batteries per unit weight, and don't discharge very quickly if you store them.

- Learn how to best take care of your batteries. Some need to be completely drained on a regular basis to have the longest life. Others should be charged at every opportunity.

- To maximize battery life, turn off power-hungry features that you don't need. In particular, turn off the LCD screen except when you absolutely must use it.

■   The memory card that came with your camera almost certainly offers less storage than you need. Consider buying a higher capacity card.

■   If your memory card fills up, you may be able to find a kiosk that will let you burn a CD and delete the images from your camera.

■   Consider bringing your notebook computer with you to give you a place to offload the photos when the memory card fills up.

■   Another possibility for freeing up the memory card is to move the files to someone else's computer, then save them on removable storage, such as a Zip disk or CD-R disc.

■   To free up relatively small memory cards—with 8 or 16 MB—it can be useful to e-mail the photos to yourself or move them to a PDA.

Part II

Getting Creative and Cutting Loose

Want to make sure your photo looks as good as it can look? Want to go beyond that and add artistic effects, like blurring the image or distorting it to make it more eye catching? How about modifying the image so it looks like a watercolor painting or a charcoal sketch? Ever thought of turning your photos into postcards? Greeting cards? Want to add frames around the pictures, join two photos together into one print, or...something else?

Well, then, you've come to the right place.

What you'll find in this part of the book are ways to dress up your photos in their Sunday best. You'll also find ways to dress them up for partying on Saturday night.

In the first category, you'll find out how to edit your pictures to eliminate minor problems like red eye, how to adjust their size and resolution to make them look better when you print them, how to adjust color if you're not happy with the way the colors look, and more.

In the second category, you'll find out how to stitch pictures together to form a panorama, and, more important, how to take the pictures so you can stitch them together successfully. You'll also learn how to make your pictures into postcards, greeting cards, and the like. In truth, you can create all sorts of special effects with digital photos, and most of them aren't even hard to do, if you know how. This part will tell you how.

Chapter 7

Getting Creative with Your Camera

Creativity is where you find it; it's not something you'll learn from a book, and we don't mean the title of this chapter to imply that we can teach you to be creative. However, we can point to some tools that will help you unleash your creativity.

Some of the most important tools are things we've covered in earlier chapters, because they apply just as much to film cameras as digital cameras. We're thinking specifically of macro mode, zoom, various options for flash, and various advanced settings. If you skipped over the first part of the book, or just skimmed it, we recommend that you go back and read at least Chapter 4, "Is That a Snapshot in Your Camera, or Did You Take a Photograph?" and the section "Common Features and How to Use Them," in Chapter 3, "Getting Started with Digital Photography." Chapter 4 is all about some rules of thumb that will help you take good pictures. The section in Chapter 3 includes a discussion of how to frame your shots, how to best use a zoom feature, choices for flash, and various advanced features.

In this chapter, we'll take a look at a special-purpose panorama format and custom definitions for clusters of settings. We'll start with panoramas.

Stitching a Panorama Together

We touched on the idea of stitching photos together to create panoramas in Chapter 2, "Knowing (and Choosing) Your Camera." In case you skipped that discussion, stitched panoramas, like the one in Figure 7-1, are panoramas that you create from literally (or at least digitally) stitching photos together using a stitching program. Many cameras come with a stitching program. For those that don't, you can buy a program separately.

Figure 7-1 We created this panorama by stitching four pictures together.

You may have tried to do something like this with film prints—taking the series of pictures with chemical film, and then, when you got the pictures back, taping them together with edges overlapping (and most often misaligned). The process is unwieldy but often serves the purpose well enough. With digital stitching, however, the process is far more sophisticated and it's easy enough that, with a little practice, you're all but guaranteed success. Of course, you can scan film into digital format, and then stitch the photos together, but if you start with a digital camera, you don't have to go through the scanning step.

Taking the picture—or pictures—is easy. Ideally, you should mount the camera on a tripod, and rotate it just enough from one picture to the next so that each picture overlaps with the one before. Some stitching programs will let you overlap rows of pictures as well, so you can have two or more rows of photos joining to create the panorama.

Many cameras offer a stitching panorama mode to help get the overlap right. For example, the Epson PhotoPC 3100Z that we used for most of the pictures in this book—and for the photos for the panorama in this section in particular—offers a mode that shows you a transparent slice of the side of the picture you have just taken. When you rotate the camera to the next position, you line up the transparent section over the same part of the scene for the next picture before taking the shot. The Epson camera also supports two rows for panoramas, and shows a transparent slice for lining up the second row to overlap with the first as well.

Tip   If you come across the most beautiful panorama you've ever seen, and don't have a tripod, use some of the techniques we discussed in Chapter 4 for holding a camera steady.

Any given camera may or may not have a stitching panorama mode, and those that have one may limit the number of pictures you can take in one series, which means you may or may not be able to take enough pictures with a panorama mode to fill out a full 360 degrees.

Just because your camera doesn't have a stitching panorama mode, or has a limited one, doesn't mean you can't take photos for a stitched panorama. It just means you'll have to note a landmark on the side (or top or bottom) of each shot yourself, then make sure you include it in the next picture to overlap the photos. The real limiting factor for stitched panoramas is the software, which controls how many pictures it can stitch together in one row and how many rows it can handle. If the software won't let you do the kinds of panoramas you want to do (or if your camera didn't come with stitching software), there's no reason you can't buy additional software.

> **Tip** If you really want to get adventurous, take a 360-degree picture. And if you want a fun photo, take each shot with the camera's self-timer and put yourself into each scene.

Using Stitching Software

The stitching software is where the magic happens. After you take the pictures and move them to your computer, open the stitching program. For this example, we'll use Sanyo's Panorama Stitcher Light, shown in Figure 7-2.

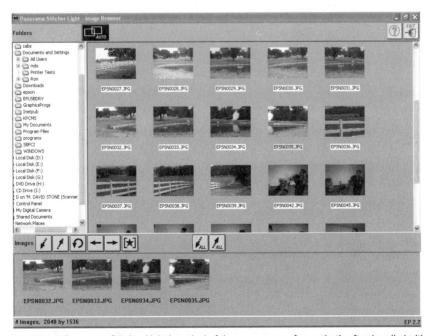

Figure 7-2 Panorama Stitcher Light is typical of the panorama software that's often bundled with cameras.

Typically, the stitching program will show you thumbnails of the photos on disk, as in the figure, and let you pick the ones you want to stitch together. You then specify the photos and the sequence. In Panorama Stitcher Light, you high-light each picture in order and import it into the program. Figure 7-2 shows the four photos for the panorama lined up at the bottom of the screen.

Getting the photos aligned properly may be a two- or three-step process, depending on the program. In Panorama Stitcher Light, you first give the com-mand to automatically align the photos, and the program tries to find the right overlapping match, as in Figure 7-3.

Figure 7-3 The program may be able to line up the photos automatically.

In this case, the automatic alignment worked well. If it doesn't, the program lets you move the pictures by dragging and dropping them to get them lined up properly. When you're happy with the alignment, you give the command to stitch the pictures together.

At that point you may wind up with an image like the one in Figure 7-4, with everything aligned, but with ragged top and bottom edges. The stitching program may provide a trim feature to let you crop the photo as needed. In Fig-ure 7-4, we've defined the area to trim, which shows as a relatively subtle dashed rectangle. (If the stitching program won't let you crop the image, you can crop it in a photo editor.)

Figure 7-4 The stitched picture may show ragged edges, which you'll want to crop.

Figure 7-5, finally, shows the finished picture—the same one we showed at the beginning of this discussion—ready for printing at high quality to hang on your wall.

Figure 7-5 The finished picture shows no seams.

Finally for this subject, here are some useful hints for taking pictures suitable for stitching together as panoramas:

- The camera's liquid crystal display (LCD) can be hard to see in sunlight. One way to overcome that problem so you can see whatever visual aid the camera gives you for stitched panoramas is to drape a piece of material over the camera and your head to form a dark viewing environment.

- Make sure your tripod is sturdy and level. Many tripods come with a leveling device. If yours has one, take advantage of it.

- Don't move the camera or the tripod during the sequence—even a little. To help avoid moving it, use the camera's self-timer to take the pictures and eliminate any chance of moving the camera accidentally while pressing the shutter button.

- Consider taking all the photos in the manual exposure mode. This will decrease the likelihood of inconsistent exposure from one photo to the next caused by variations imposed by automatic exposure.

- When you import the photos into the stitching program, make sure you line them up in the right order for the software to stitch them.

Storing and Using Clusters of Settings

Warning   Do not skip this section, or you'll miss some important stuff.

We included that warning as the first thing in this section because we know that a percentage of readers (you know who you are) will be ready to jump over this section as soon as they realize what it's about. Those are precisely the people who most need to read this. We urge you to read it even if you know all about stored clusters of settings and hate the idea—no…make that *especially* if you hate the idea. Read it even if your camera doesn't offer this feature. You may decide it's worth getting on your next camera.

Having said all that, we'll start with some comments for the people who weren't about to jump over the section.

The Logic of Clusters

One of the features we covered in Chapter 3, "Getting Started with Digital Photography," in the section "Sets of Settings," was, well, sets of settings. Not all cameras offer this feature, and those that do use their own names for it—like *scenes*, or *programmed settings*. By whatever name, however, the idea is to store a set of camera settings that you'll always want to use for certain types of pictures. The photos in Figure 7-6, for example, are different kinds of photos, and each one needs a different set of settings.

Figure 7-6 For different types of pictures, like the two shown here, you need different settings.

For portrait shots, like the one on the left side of the figure, you might always want to use an aperture priority mode with a large opening to ensure the background is out of focus, forced flash to minimize shadows and light up a backlit subject, and spot metering (taking readings of a spot at the center of the scene rather than over the entire scene as viewed by the lens).

For landscapes, like the photo on the right, you might always want to use aperture priority (because you're more interested in controlling depth of field than in stopping action) and matrix metering (taking readings of the entire scene). You may also want to turn the flash off so it doesn't light up the fore-ground. By storing each of these clusters of settings under a single name—we'll call each cluster of settings a mode—you can change all the settings at once by picking a mode.

Preprogrammed modes like these (*preprogrammed* because they are set by the camera manufacturer) tend to get ignored. Point-and-shoot photographers, who could most benefit from them, often don't understand how to use them. Mildly creative photographers tend to be frustrated by not knowing what the camera settings are for each mode. Prosumers tend to look at the prepro-grammed modes with something approaching contempt: real photographers change their own settings.

Clusters of Settings as a Power User's Tool

We'd argue that stored settings can be extremely useful. An analogy may help explain why.

Many computer programs offer a feature called macros—which, in case you're not familiar with the term, is a completely different thing from a macro mode in photography. On a computer, a macro feature provides do-it-yourself automation. Macros are actually programs, but you don't have to be a programmer to create one. You can simply record the steps you want to automate.

For example, in writing this book, there are times when we want to keep some text we've written from showing on screen. We might have a paragraph that we're considering deleting, say, but haven't made a final decision about. The program we're using for writing lets us format the text as hidden so it doesn't show. There's a minor problem with doing that, however. Because the text is hidden, it's easy to delete it accidentally when you delete something near it.

Our solution is to mark the hidden text by putting left and right angle brackets around it, so if we see <> we know there's some hidden text there. But that's still easy to miss, so we also format the angle brackets in red to stand out better. The entire procedure takes 24 keystrokes. (You can do the same thing in fewer mouse clicks, but then you have to keep moving between the mouse and the keyboard, which slows things down.) By recording the procedure in a macro and then defining the macro to run with a single keystroke, we can mark the text as hidden, insert the brackets, and format the brackets in red all with a single keystroke.

That's a macro. In the world of computers, it's a power user's tool.

Preprogrammed modes in cameras are much like macros in spirit. They automate what you want to do to let you get it done more quickly. And if you're a prosumer (or prosumer wanna-be) who isn't using the feature because you think it's for wimps, you're missing the point.

Unfortunately, there are some problems with preprogrammed modes. For mildly creative photographers and prosumers, they're truly useful only if you know what settings are actually stored in each programmed setting, and, more important, they're useful only if those settings are what you actually want to use for the particular kind of shot.

Not knowing what the settings are undermines the usefulness of the modes—and too many camera manuals fail to give that information. If you don't know what the settings are, you can't be sure you want to use them. And, of course, if you know what the settings are, but disagree with them, you know for sure that you don't want to use them. It's not unusual for a manufacturer to define an indoor mode with the camera set to flash with red-eye reduction. If you agree with our recommendation not to use red-eye reduction, you won't want to use that mode.

Define Your Own Clusters

The good news is that some cameras let you store your own modes, using whatever settings you like. For these cameras, the analogy to macros in computer programs is complete. You get to define exactly the cluster of settings you want and automate the process of changing the settings on your camera through a custom definition.

Alas, if your camera doesn't offer this feature, there's no way to add it—unless you buy a new camera. But if it does offer the feature, we hope we've convinced you that it's worth taking advantage of, especially if you're a prosumer who hasn't thought of it as a power user's tool until now.

In any case, for those whose cameras offer the ability to define their own modes, here are some suggestions for settings you can use for some common kinds of photos. This is just a starter set, and the suggestions are just that—suggestions. You may want to—or may have to—modify them to meet your preferences or your camera's features. If your camera doesn't have an aperture priority setting, for example, you won't be able to set the camera to use aperture priority, as suggested for outdoor portraits. With that in mind, we've included the reasons for each setting. If you understand why we picked each one, you can apply the same reasoning to the settings actually available in your camera.

Outdoor Portrait

Settings:

- **Forced flash**   To minimize shadows and light up a backlit subject.

- **Aperture priority**   There is little or no need for a fast shutter speed, and you may want to control the depth of field.

- **Spot metering**   The subject of the photo is in a small area—a person's face.

Landscape

Settings:

- **Flash set to none**   Flash lights a scene only up to 10 to 15 feet away.

- **Aperture priority**   Use a small aperture to make sure both foreground and background are in focus.

- **Matrix metering**   You'll need to base the auto-exposure on the whole scene.

Action in Daylight

Settings:

- **Flash set to none**   Daylight is all you need for lighting.

- **Shutter speed priority**   You'll need a fast shutter speed to stop action.

- **Matrix metering**   You'll need to set the exposure for the full scene of the action.

Sunset

Settings:

- **Flash set to none**   You want to see the sunset, not the foreground.

- **Full automatic mode**   Let the camera determine shutter speed and aperture settings.

- **Matrix metering**   Gives you great colors in the sunset.

- **Manual focus set to infinity**   Ensures that you're focused on the sunset and not something closer.

- **White balance**   Set to roughly 5000 to 5500 degrees Kelvin, if your camera will let you, to bring out reds and yellows.

Fireworks

Settings:

- **Flash set to none**   The fireworks won't benefit from the extra light.

- **Shutter speed priority**   Use a slow shutter speed to capture bursts.

- **Manual focus set to infinity**   An auto-focus setting will have problems finding something to focus on.

Key Points

- Some cameras offer a panorama mode that will help you take pictures suitable for stitching together. For example, the camera may show you a slice of the previous image with each step, so you can line up the next picture.

- If your camera doesn't have a stitched panorama mode, you can still take pictures for stitching into a panorama; you just have to determine how much to overlap the shots yourself, and be sure that you overlap them properly.

■   Your stitching software determines the limitations, if any, for stitching pictures together. It may allow any number of pictures in a row, and any arbitrary number of rows.

■   When you take photos for stitching together, it's best to use a tripod. If you don't have one available, pay particular attention to holding the camera steady, and taking pictures in a single horizontal sweep, so they will stitch together well.

■   Programmed clusters of settings for particular kinds of photos are best understood as the equivalent to a macro in a computer program—a way to automate a task that would otherwise take more steps.

■   Preprogrammed clusters of settings—the ones that you may find already programmed in your camera—can be problematic, because the camera manufacturer may not have chosen the settings you would choose, and there is often no way to tell what the settings are.

■   Some cameras let you define your own clusters of settings. This is truly a power user's tool, since it lets you define the settings you want to use in particular situations and set them all with a single command.

Chapter 8

Fun with Pictures: Basic Editing

With film photography, most people's creative involvement with their photos stops with clicking the shutter. At some point you'll drop the film off for developing, get the developed photos back, look at the results, share them with friends, put them in albums, and so on, but you probably won't do anything further to the photos themselves.

With digital photography, the equivalent moment to dropping your film off for developing is moving the files to your computer; and if you want to, you can simply print out the photos and treat them the way you would treat film photos. But that would be a shame, because with digital photography, this is the point where the creative fun can really begin.

Taking photos and standing by the results is fine, but why should you when you have a tool handy that can improve the results—and even let you add special effects, as in Figure 8-1. That's what graphics editors let you do. And most are so easy to use (at least for basic editing) that there's no good reason not to try. Here then, is an introduction to basic photo editing, or if you prefer something that sounds less formal, here are some easy ways to improve your pictures with little or no work.

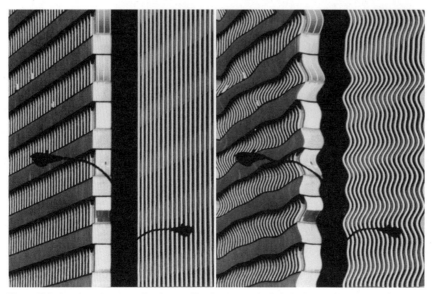

Figure 8-1 A camera lets you take pictures like the one on the left. A photo editor lets you fix problems or add interesting effects, as with the version on the right.

What's a Photo Editor and How to Get One

Odds are you already have a photo editor, and quite possibly more than one. Most cameras come with a photo editor in the package. So do most scanners. So do some programs that you might not expect would have a photo editor—notably Microsoft Office, which includes one with the straightforward name Microsoft Photo Editor (just in case you have any doubts about what it's for).

We've collected a bunch of photo editors over the years and a few more came with the cameras we gathered as aids in writing this book. We'll use several of them to illustrate the things we're talking about in this chapter and in Chapter 9, "Advanced Editing: Fixing Flawed Photos," and Chapter 10, "More Fun with Pictures: Special-Purpose Editing." Before we get to any photo editing, however, you might find it helpful to know what makes a program a photo editor.

Types of Graphic Programs

Graphics fall into two broad categories: vector graphics and bitmap graphics. We'll tackle vector graphics first.

Vector graphics define lines and shapes as actual objects. Draw a straight line, for example, and a vector graphics program knows where it starts and where it stops. Draw a circle, and the program knows where the center is and how large the radius is. Add a little more information about color and the thickness of the lines, and you can exactly describe the line or shape.

Lingo   *Vector graphics* store lines and shapes by describing them as actual objects. A line, for example, can be defined by the position of the two end points, color, and thickness.

Images, no matter how complex, ultimately can be broken down into combinations of tiny lines and shapes, and computers can store the images as a collection of definitions for those lines and shapes. This has some advantages for some situations (which we won't bother talking about here, because they have no relevance to photography).

By convention, programs that use vector graphics are called *drawing programs*. If you can remember that, then when you see a program that calls itself a drawing program, or includes the word *draw* in its name, or you see an option like the *drawing* toolbar in Microsoft Word, you'll know immediately that it's a vector graphics program or feature, rather than a photo editor.

Lingo   Graphics programs that use vector graphics are called *drawing programs*.

The alternative to describing images as a collection of lines and shapes is to think about the image as a grid of bits (in the sense of being the smallest part of the image, not a computer bit). You then define the position and color of each bit. Bitmapped graphics programs don't know anything about lines or circles or any other shape as individual objects. They only know about the bits at each position in the grid. If enough bits of the same color lie in a line along the grid, and they're surrounded by bits of other colors, you get a line. If they lie in a position that describes a circle, you get a circle.

When you describe an image like this, each bit is mapped to a specific location on the grid, which is to say, the image consists of a bunch of mapped bits, and the image is called *bitmapped*. Programs that deal in this kind of graphics are known as *bitmapped graphics programs*.

Lingo   *Bitmapped graphics* define an image pixel by pixel, or bit by bit, mapping each pixel to a specific location.

Figure 8-2 shows a graphic that we created in a bitmapped program. We then zoomed in on the image far enough so you can see the individual bits in the grid, and we set the program to show the grid.

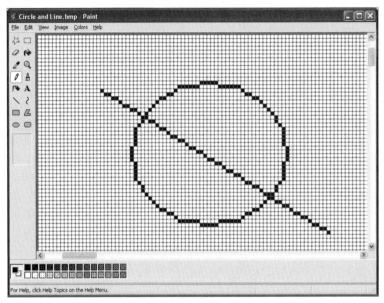

Figure 8-2 Zoom in on (or enlarge) a bitmapped graphic far enough, and you can see the individual pixels, or bits.

As you can see in the figure, you won't be far off if you think of bitmapped graphics as drawn on a piece of graph paper. The individual cells in the grid are filled in according to the definition for each bit. It's the computer equivalent of a paint-by-numbers kit.

By convention, bitmapped graphics programs are called *paint programs*. So just as the word *draw* in a program name tells you that you're dealing with a vector graphics program, the word *paint* tells you that you're dealing with a bit-mapped program.

Lingo   Graphics programs that use bitmapped graphics are called *paint programs*.

If you have a typical installation of Microsoft Windows on your computer, you have at least one bitmapped graphics program, called Microsoft Paint—which is the program we used for Figure 8-2. If you're not familiar with it, you can find it by opening the Start menu and choosing All Programs or Programs, depending on your version of Windows, and then Accessories. Paint will be one of the menu items. Figure 8-2 shows the Microsoft Windows XP Professional version of Paint. Other versions of Windows have slightly different versions of the program.

Photo Editors

You may have noticed that this discussion of bits in a bitmapped program sounds a lot like earlier discussions in this book about pixels in a photograph.

That's because the two are identical. Photos are bitmapped images, and the pixels are the bits, which means you can view a photo in any bitmapped graphics program. You can even edit photos in any bitmapped graphics program to do things like crop the image. However, some programs go beyond graphics editing to offer special features specifically designed for photos—removing red eye, for example, or adding special effects. Most make the difference clear by including the word *photo* somewhere in their names.

Choosing a Photo Editor

As with most software, choosing the right photo editor is largely a matter of individual taste. A heavyweight in this category is Adobe Photoshop. It is aimed at professional photographers and graphic artists, and has a fairly steep learning curve.

If you get serious about editing your photos, you might want to get one of these professional-level programs. However the photo editors that typically come with cameras and scanners aimed at nonprofessionals are designed to be far easier to use. Alas, some of them are extremely limited, but others provide more than enough features for most people. Many of them even provide special effects that you won't find in Photoshop.

One common shortcoming in many photo editors for nonprofessionals is that they leave relatively little room for the photo on screen. Figure 8-3, for example, shows a photo in ArcSoft's PhotoImpression.

Figure 8-3 Some photo editors, like PhotoImpression, reserve much of the screen for the program controls.

PhotoImpression is a capable editor, which is one of the reasons we chose it for this example. But in its effort to make the commands easy to find and use—which it does well—it leaves only a little more than half of the screen for the photo. The image in Figure 8-3 was captured at 800 × 600 screen resolution. The area for the photo is roughly 550 × 360 pixels.

You can find programs that devote more of the screen to the photo, and programs that use even less for the photo. We prefer programs that let us use as much of the screen as possible, so we can show the photo at a reasonably large size, and see more of it at once as we edit it. Photoshop, for example, lets you use almost the entire screen while editing, as shown in Figure 8-4.

Figure 8-4 Some photo editors, like Photoshop, reserve most of the screen for the photo.

Among programs that are designed for nonprofessionals, Jasc's After Shot Premium Edition (previously Image Expert), Adobe Photoshop Elements, and Microsoft Picture It! Digital Image Pro 7.0 are noteworthy for leaving most of the screen for the photo.

Photo Editing

In the rest of this chapter, and for the next two chapters, we'll be giving you a quick introduction to using a photo editor to improve your photos. We'll be using several programs in our examples, and will mention the particular programs as appropriate. However, our goal here is not to teach you how to use any particular piece of software. We're trying to show you the kinds of things you can do with photo editors in general and give you an introduction to how to edit photos. Details, and even the editing features available, will vary depending on the program you have.

This chapter concentrates on features that you can find in just about any photo editor, but it's conceivable that even some of these features may not be available in the editor you have. Keep that in mind as you look at these examples, and then look for the right commands in your program.

One other note: you'll probably recognize some of the photos in the examples from elsewhere in this book. In many cases we had to edit photos before we could use them—to fix flaws in lighting, crop the image, change resolution, remove unwanted elements that we couldn't control when taking the picture, or otherwise turn a photo from what it was into what we needed it to be. We thought you might find it interesting to see how we edited some of the photos. We also thought it would help drive home the point about how useful photo editing can be.

Rotating an Image

The most common reason for rotating an image is that you shot it in vertical format. If you did that, it will show sideways when you open it up in your photo editor, as in Figure 8-5.

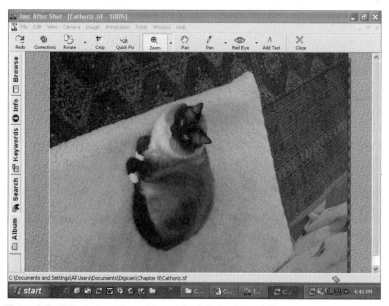

Figure 8-5 Vertical shots show sideways by default when you first open them (in this case, in After Shot Premium Edition).

Any photo editor will let you rotate the image. Almost all have simple menu choices for rotating 90 degrees clockwise or counterclockwise. Many offer a choice of 180 degrees as well. A few don't offer a menu choice, but will let you go into a Rotate mode, then use your mouse to rotate the picture. This typically involves *dragging and dropping* a corner of the picture, meaning that you click on the corner of the photo and, while holding the mouse button down, *drag* the corner to indicate how far to rotate the picture. When you're done, you release the mouse button to *drop* the picture, which indicates that you're finished dragging.

Lingo   *Drag and drop* is a mouse-based technique. To *drag* an object, you click on it and, while holding the mouse button down, drag it to a new location. To *drop* it in place, you release the mouse button.

In Figure 8-6, we rotated the image 90 degrees counterclockwise. Once you've rotated the image to the right orientation, it's a lot easier to work with it for further editing.

Figure 8-6 You can rotate the picture, and save it in the new orientation.

We'll come back to this photo in a moment; there's lots more to do with it. First, we need to mention another reason you might want to rotate an image.

Rotating to Reframe an Image

In Chapter 4, in the section "Quick Rules," we talked about the importance of lining up a horizontal or vertical element in a shot with the sides or the top and bottom edges of the picture. You may remember the bird feeder at the top of Figure 8-7 from Chapter 4, where we pointed out that it should be hanging straight up and down. You can fix this problem in many, but not all, editors. Look for a feature that lets you drag and drop the image, or one that lets you rotate it by some exact number of degrees, to wind up with a fixed picture, as at the bottom of the figure.

Figure 8-7 The bird feeder shown on top looks odd because it's not parallel to the edge of the frame. The version on the bottom is the same shot, rotated by just 5 degrees.

Cropping to Clean Up Clutter

There are all sorts of reasons why you may want to crop a photo. One of the most common is that there is some distracting element on the left, right, top, or bottom that you want to get rid of. Another is that you want to close in on the subject of the photo. For an example, take a look at Figure 8-8. This is the same photo we rotated to a vertical position before, but we've now zoomed out so you can see most of the photo.

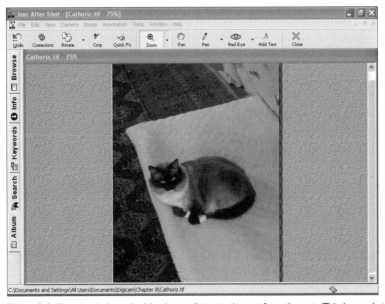

Figure 8-8 The visual clutter in this picture distracts the eye from the cat. (This is nearly the entire photo.)

This photo is a cluttered mess. The subject is the cat. The rug and the slice of floor and blanket in the upper right corner are distracting. Even if they weren't a problem, the cat is taking up only a small portion of the picture, and should surely be larger relative to the picture size. Ideally, you would have noticed all this when you took the picture, and framed the scene differently, but in the real world, you might not have been able to get close enough to the cat before she ran away. Realistically, you have to deal with what you've got.

Every photo editor has some sort of cropping tool. Typically, you'll choose the cropping tool or choose a selection tool from a toolbox, click on one corner of the cropped image you want to define with the mouse, and then drag the mouse across the image to select the part of the image you want to keep, as shown in Figure 8-9.

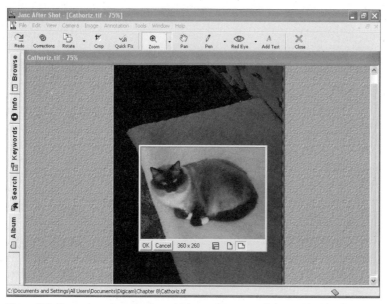

Figure 8-9 The rectangular selection, which stands out as a light area in After Shot, shows the part of the picture you'll keep after you give the command to crop.

When you're happy with the selection, you can give the command to crop the image. In Figure 8-10, we've both cropped the photo and zoomed in on it so you can see it better.

Figure 8-10 After cropping, the cat takes up most of the photo.

This is a much better picture. The cat is dominating the photograph, and there's much less clutter. We're not done improving it yet. (For one thing, we want to get rid of that distracting oriental rug entirely; for another, the cat is beginning to show some pixelization when we look at it at full-screen size.) But again, before we go any further with this shot, let's go on a short side trip, and discuss another reason why you might want to crop an image.

Cropping Can Make a Boring Shot More Interesting

You can also crop a picture with an eye toward making it more interesting. If you took a photo with the subject dead center, for example, and later decide it's a dull picture, you may be able to shift it somewhat to turn it into a more interesting photograph.

Figure 8-11, for example, shows the same shot before and after cropping. We'd argue that the *after* version, which follows the rule of thirds, is a more interesting photo. (We covered the rule of thirds in Chapter 4, "Is That a Snapshot in Your Camera, or Did You Take a Photograph?")

Figure 8-11 The photo on the top is the original version. The one on the bottom is the same photo after cropping.

Serious photographers will tell you that you should take the shot right in the first place. But, as a practical matter, cropping gives you what lawyers would call a second bite at the apple. We'll take a second bite too, and urge that if you skipped over Chapter 4, you go back to it and look at the section "Choosing a Composition." The discussion there applies just as much to cropping as to taking pictures.

Flipping

Just about any photo editor will include a command to let you flip pictures horizontally to give you a mirror image of the picture. If you're printing photos as stand-alone photos, this isn't a particularly useful feature. But if you're using the photos in other formats—inserting them in a newsletter or Web page, for example, it can be a valuable tool.

Suppose that you're writing an end-of-year newsletter for friends and family, and want to include pictures of your dog and cat next to each other, as in Figure 8-12.

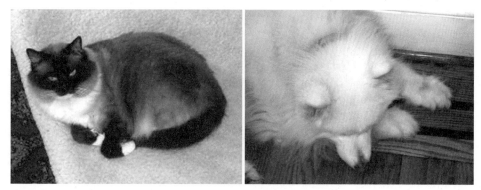

Figure 8-12 Our ubiquitous cat, and a dog, facing different ways.

Unfortunately, the pictures don't go together very well with the animals facing away from each other. You could try switching them, as in Figure 8-13, but that doesn't work very well visually either.

Figure 8-13 The two pictures don't work well together this way either.

However, if you flip one of the pictures, as in Figure 8-14, so both the dog and cat are facing the same way, the problem is solved, and the two pictures work together. Note that most programs will let you flip a photo horizontally, as we've flipped the picture here, or vertically. If you flip a picture vertically, and then rotate it 180 degrees, the result is the same as if you flip it horizontally.

Figure 8-14 Flip one of the pictures so the dog and cat are both facing the same way, and they look much better together.

Size and Resolution

There's nothing inherent in a photo that forces it to be a given size. However any photo editor will let you define a size for the photo to print. If you're not particular about how the photo looks, you can simply set the size and let it go at that. Just find out where you change the size in the program—in a size dialog box, in a text box when you print, or elsewhere.

Unfortunately, there's a small complication. When you change the size you may change the quality of the printed output. Say you have a photo that's 800 pixels across. If you print it at 4 inches across, you'll get 200 pixels per inch (ppi), which will give you reasonably good output quality. Print it at 10 inches across, however, and you'll get 80 ppi, which won't give you the same quality. Resample the image to 200 ppi at the new size, and you'll gain a (usually) small, but noticeable increase in quality.

It's this relationship between size and resolution that you need to be concerned about. The two are linked in a way that can be confusing. But it doesn't have to be. Read on.

Sorting Out Some Tangled Threads

The secret to knowing your way around size and resolution is to compartmentalize your thinking. First, understand that resizing an image for viewing on screen is a completely different issue than resizing for printing. Keep the two firmly separated in your mind, each in its own pigeon hole. We'll talk about resizing for printing first, and then come back to the issue of resizing for viewing on screen later.

Second, don't try to change size and resolution at the same time, if you can avoid it. Some programs will let you do that, but until you have a firm grasp of the relationship between the two, we advise against it. It's better to adjust the size and resolution in two steps, and go through the process without a mistake, rather than try to do both at once and have to start over.

There are two basic variations on how to set size and resolution, depending on the photo editor you're using. Virtually any photo editor will include a command called Size, Resize, or something similar. Alas, it doesn't always mean the same thing. In some editors, the command lets you specify a size for printing the photo— 4 × 6 inches, 8 × 10 inches, or whatever you like. You can then find another option for changing resolution. In other editors, however, the size command changes the photo's size in pixels —what we've called pixel resolution in this book. That's really a resolution setting rather than a size setting. Typically in those programs you specify a size to print the photo when you give the Print command.

Most high-end photo editors aimed at professionals and serious amateurs use the first approach. Most low-end photo editors—the kind you most likely got with your camera—use the second approach. Fortunately for people who have to switch back and forth between the two, there really isn't all that much difference conceptually in the procedure you have to follow for setting size and resolution with one approach or the other. To keep from getting confused in either kind of program, however, it's essential that you understand what happens when you *resample* a photograph to change the number of pixels. It also helps to know why you would want to.

Lingo   When you *resample* a photograph, you change the number of pixels in it.

Sampling, Resampling, and Resolution

When you take a photo, your camera's sensor doesn't simply record the image coming though the lens. The technically correct description for what it's doing is called *sampling*. The camera divides the scene into a grid, with so many cells across and so many cells down, based on the resolution you told it to use. It then essentially looks at each cell as a single, indivisible unit, or sample, of the image. Each sample becomes a pixel in the photograph. Set the resolution higher, and the camera takes more samples; set it lower, and it takes fewer samples.

Once you get the photo onto your computer, you may decide that you want a different resolution—which means a different number of samples. As we've discussed elsewhere in this book, if the resolution is too low—with too few samples—you'll see pixelization. If the resolution is too high—with too many samples—you may not be able to see the photo at its best on screen, or you may be wasting disk space on extra information that won't improve the quality of your printed photo. In either of these cases, you'll want to change the resolution, which means you have to resample the photo.

There are different ways that a program can resample an image. To resample to a lower resolution, for example, it could simply throw out some pixels. However, a better approach is for it to examine the pixels in the original image and replace them with new pixels that effectively split the difference in color between two or more pixels in the old image. Similarly, to resample to a higher resolution, a program can simply add pixels by duplicating some that are already there, or it can create new pixels by looking at the pixels that are nearby the new pixel, and, once again, effectively splitting the difference.

Whether adding or subtracting pixels, the process of splitting the difference, or interpolating pixels, is the more sophisticated approach, and we can't imagine anyone creating a photo editor that doesn't use some variation on interpolation. In high-end programs, you even have a choice of which variation to use. Photoshop and Photoshop Elements, for example, give you three choices. (If you have more than one choice, check the program manual or help file to find out the differences, and why you might want to use one rather than another. In Photoshop and Photoshop Elements, bicubic is generally considered the best choice.)

The point is that when you change the pixel resolution of a photo, you're actually resampling the image, and that's true no matter what the command is called. Armed with that information, let's get back to talking about how to change size and resolution.

Size and Resize Your Photos as Needed

Remember the watchword for resizing and resampling photos: compartmentalization. We already suggested that you should set size separately from resolution. You should also compartmentalize the way you think about resolution.

As we've discussed elsewhere in this book, there are two ways to talk about the resolution of a photo: the pixel resolution, which tells you how many pixels are in the photo, and pixels per inch (ppi). The two are related by the size photo you plan to print. The pixel resolution divided by the number of inches across or down tells you the number of ppi. Alternatively, the ppi setting times the size for printing tells you the number of pixels.

You can avoid all sorts of confusion when you're resizing and resampling your photos by focusing on one kind of resolution or the other, and trying not to go back and forth between the two any more than you must. In fact, it's usually most convenient to think in terms of ppi, if your program allows it, and leave the pixel resolution for the program to figure out. If you have to deal with both because of the way your program requires you to enter the information, do the conversion when necessary, but stay with a single approach as much as possible. Remember, you'll get to the same result no matter which approach you take.

Keeping all this in mind (but in different pigeon holes), here's a strategy for setting size and resolution without getting confused:

- First, decide how large you plan to print the photo. That's the critical piece of information you need before you can do anything else. Don't skip over this step, even if the program you're using won't let you set print size until you print.

- Second, if your photo editor lets you define a size for the photo separately from printing the photo, as in the Photoshop Image Size dialog box shown in Figure 8-15, set the size.

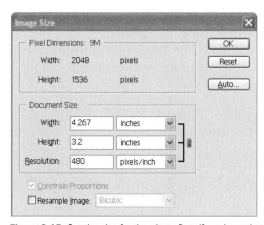

Figure 8-15   Set the size for the photo first, if you have that option.

- Be careful not to let the program resample the photo at this step. Note that in the figure, we've cleared the Resample Image check box. This tells the program to resize the image without resampling. The pixel resolution will stay the same, and Photoshop will recalculate the ppi setting, based on the new size. If you set just the width or just the height, most programs will automatically set the other to maintain the same proportions—otherwise you'll distort the image. (Your program may refer to the proportions as the aspect ratio.)

- After you set the size, be sure to accept the new size before moving on to the next step, assuming that the program allows that. This may mean choosing an Apply button or choosing the OK button. You might even want to save the file at this point. Having said that, don't let it throw you if the program doesn't let you accept the new size, but forces you to set a resolution also and then resample. Digital Image Pro, for example, makes you do both at once, by way of numbered steps.

■ Next, determine the resolution you want to use in ppi. We recommend at least 150 ppi. A higher resolution may improve the image, but keep in mind that it's hard to see any improvement beyond 200 ppi, and anything more than 300 ppi is just using up disk space without improving image quality.

■ If your program requires that you set resolution in ppi, or it gives you the choice, enter the ppi setting in the appropriate place in the program, and let the program resample the image. And that's it. You're done. Don't even think about pixel resolution.

■ If your program requires that you set resolution as the number of pixels in the image, you'll have to calculate it. Multiply either the height or width of the image by the number of pixels you want per inch, and enter the result in the program. Whichever setting you enter—width or height—the program should automatically recalculate the other. If it doesn't, do the calculation yourself and enter the result. Then let the program resample the image.

At this point, you're ready to print. If the program let you set the print size earlier, simply give the print command. If it makes you set the size as part of the process of printing, give the command to print, and don't forget to enter the size before actually printing.

Resizing for the Screen

Resizing photos for viewing on screen is much easier than resizing them for printing, because when you're talking about the screen, resolution and size are essentially the same thing. At least that's true in the sense that for any given screen resolution, the photo's pixel resolution tells you how much of the screen it will take up if you don't zoom in or zoom out on the photo.

As we've pointed out elsewhere in this book, for best viewing, you want to show a photo on screen at 100 percent, meaning one pixel on the screen gets used for each pixel in the photo. To do that and still see the whole photo, you need a pixel resolution for the photo that's less than the resolution for the screen you expect to look at it on. So to resize or resample a photo for the screen, the only thing you need to know is what screen resolution you're expecting to use and how much of the screen will be devoted to the photo. A good rule of thumb that we mentioned in an earlier chapter is to use roughly the next lower standard screen resolution. So for viewing on a 800 × 600 pixel screen, you'll want to resample the photo to 640 × 480 or so.

Resizing—or resampling—for viewing on screen, then, is a two step procedure. First decide on the pixel resolution you want for the photo, and then resample the photo to that resolution. It's that simple.

Oh, one other thing: ignore references to ppi. You'll often hear that screen resolution is 72 or 96 ppi, but that's just a convenient fiction. For any given screen resolution, the actual number of pixels per inch will vary depending on the physical size of the screen. At 800 × 600 resolution, say, the ppi on a 17-inch monitor won't be the same as on a 21-inch monitor.

About Cropping and Resampling

Before we leave the subject of resampling, we should also point out that you'll often need to resample after cropping a photo. Remember the cat we've been using for illustrating much of this chapter. When we last cropped in on it, we mentioned that it was beginning to show some pixelization on screen. That's because the image, which was relatively low resolution to begin with, at 640 × 480 resolution, had reached the point of being only 290 pixels wide. View it at full size on screen, or print it at 8 × 10 inches, and the pixelization is annoyingly obvious, as the enlargement in Figure 8-16 shows.

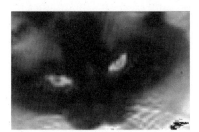

Figure 8-16 Crop in on a photo enough, and it will show pixelization.

The fix is to resample the image. Change the resolution to, say, 2000 pixels across, and you can print it at 8 × 10 inches or show it full screen without seeing pixelization, as in the enlargement in Figure 8-17.

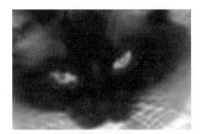

Figure 8-17 Resampling won't improve resolution, but it will get rid of pixelization.

As we discussed briefly in Chapter 1, in the section "Pixel Resolution Versus the Ability to Resolve Detail," the result may be a little blurry, but it won't be pixelized. If you don't enlarge the image all that much, the blurriness—or lack of resolution—won't be an issue.

Of course, we still have the oriental rug showing in this picture as a distraction. We'll deal with that in the next chapter, when we dip our toes into the deeper waters of advanced photo editing.

Key Points

- Graphics editors can be drawing programs, which use vector graphics that define objects by their characteristics (a line has a starting point and a length in some direction), or paint programs, which use bitmapped graphics (a line is a series of bits of a given color). Photos are bitmapped, and photo editors are bitmapped graphics programs.

- You can use the rotation feature to rotate an image that's slightly skewed, so a vertical or horizontal element lines up with the edge of the picture.

- You can use a cropping feature to cut out distracting elements, zoom in on the subject of the photo, and move the subject to a different spot relative to the center, to make a more interesting composition.

- You can flip photos horizontally to get a mirror image. This can be useful when you're inserting two photos side by side in a document, and the images aren't working together, because one's facing right, for example, and the other is facing left.

- To change resolution, you have to resample the photo, which means changing the number of samples (pixels) in the picture.

- The steps you have to go through for resampling and resizing are different for printers than for viewing on screen. When resampling for a printer, it's most convenient to think about the resolution per inch on the printed page. When resampling for a screen, you have to think about the total number of pixels in the photo's height or width.

Chapter 9

Advanced Editing: Fixing Flawed Photos

If you worked your way through Chapter 8, "Fun with Pictures: Basic Editing," you've already seen some ways to improve your photos—by cropping, rotating, flipping, and changing size and resolution. That's barely scratching the surface of what you can do. In this chapter we'll get into the fun stuff, where you have to roll up your sleeves and get some digital dirt under your fingernails.

Think about pictures you've taken that you weren't quite happy with. Better yet, look at some. Is the picture too dark or too light? Can it use more contrast—or less? Do you want to lighten or darken just a particular area? Is the color off—a little greenish, perhaps? Do you need to get rid of red eye? Is there a car sitting in the middle of a rustic scene ruining the shot? How about a person that you'd rather not be reminded of—or someone who wandered into the scene when you were taking the picture?

All of these flaws—and a lot more—are fixable most of the time; you just have to know how. By the time you finish this chapter, you will.

Techniques for Fixing Common Flaws

The first thing you need to understand about editing photos is that there's often more than one way to get to the same result. One of the issues we'll be discussing in this chapter, for example, is how to edit a picture to make unwanted objects disappear. There happen to be several ways to do that: you can copy a relatively large area from another part of the image and paste it over the part that you want to get rid of; you can use a cloning tool to copy a relatively few pixels at a time; if the spot is small enough you can use a smudge tool to make it go away, just like you would rub away a real spot on a table with your finger; and so on.

The point is that there is often more than one way to get things done. More important for our purposes, even if the program you have doesn't offer exactly the tools that we talk about here, it may still have a way to get the same result. As we talk about particular problems, we'll discuss alternative techniques for solving them, but we can't hope to cover every variation you'll find in every program. If the photo editor you have handy doesn't have the tools we mention, look for alternatives. Some programs are too limited to do everything we talk about in this chapter, but most will be able to do most of what we cover.

Red Eye

Red eye is a good example of a feature that different programs approach in different ways. Most photo editors offer a simple way to get rid of red eye, but virtually every program seems to have invented its own way.

The red in red eye is primarily in the pupils, because that's where the reflected light comes out from the eye. Since pupils normally look black, getting rid of red eye is just a matter of replacing the red with black. In some programs, you do that by zooming in on the eyes, choosing the Red Eye command, selecting a small area around the eyes, and giving the command to reduce red eye. The program finds any red inside the selected area, and turns it all into black.

A variation on this approach lets you designate the area to fix by clicking on the red areas to leave two markers—one on each eye—as in Figure 9-1 (which shows the red eye feature in Microsoft Picture It! Digital Image Pro 7.0). With the areas defined, you can give the command to fix red eye. Here again, if all goes well, the program replaces the red with black.

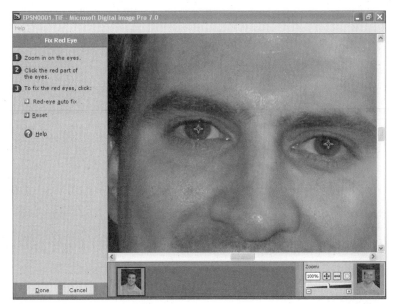

Figure 9-1 In some programs, you can target the area to fix for red eye by selecting a small area around each eye or by clicking on the area to leave a marker.

In still other programs, you choose the Red Eye command, then click on the red in the eye to instantly replace it with black. In yet others, you choose the Red Eye command, and then hold the mouse button down as you move the mouse cursor over the red in the eye. As you move the cursor back and forth, the program replaces the red with black.

If your program doesn't provide a Red Eye command, or the feature doesn't get rid of all the red, you can often zoom in on the eyes, use a selection tool to select the pupils, and fill the selection with black. (Look for a selection tool that will let you select a circle, to define the entire pupil. Then look for a Fill command that will fill the selection with a specific color. Choose black as the color for the fill.)

If all else fails, look for a paint tool that will let you zoom in on the pupils and paint them a different color, using a brush or similar feature. This sort of freehand painting takes longer than automatic red-eye tools, but it works just as well.

Adjusting Color

Sometimes, the colors you wind up with in a photo don't look quite right. We've seen photos come out a little yellowish, greenish, or shifted in other directions. Sometimes the color shift shows both on screen and in output from the printer. Sometimes it shows in the printer output only. Most often, the problem is not that the camera, monitor, or printer are doing something wrong, but simply that they don't agree on how to describe colors. Whatever the cause of a given color shift, however, most photo editing programs give you a way to fix it by adjusting the colors in the photo.

Adjusting colors is a tricky topic to talk about. There are any number of different approaches to adjusting colors—and you'll often find more than one way available in any given program. Alas, few of them are easy to understand without a lot of hands-on experience adjusting the settings and seeing the effects. For example, one of the most common approaches lets you adjust the amount of red, green, and blue in the image, or the amount of cyan, yellow, and magenta using sliders, as shown in Figure 9-2.

Figure 9-2 Some color adjustment features, like this one in Olympus Camedia Master, adjust red, green, and blue levels.

Unfortunately, this is an almost useless feature to most people, even with a preview of the photo that changes as you change the settings (which makes it a good thing that this particular program also offers other ways to adjust color.) Changing the proportions of red, green, and blue is how cameras and monitors handle colors; changing the proportions of cyan, yellow, and magenta is how printers handle colors. Neither has any relationship to the way people perceive colors. For example, given sliders for red, green, and blue, it probably would not occur to you that if grass looks too yellow in a photo, you can fix it by cutting down the amount of red and green.

A more useful approach lets you adjust the hue, saturation, and brightness (or some variation on this approach). *Hue* is what most people would think of as the color—red, orange, yellow, green, blue, and so on. *Saturation* is the intensity of the color. A pale pink isn't as saturated as a red rose. *Brightness* is the brilliance of a color. A yellow daisy is brighter than a yellow mustard color. For that matter, French's yellow mustard is a brighter yellow than Gulden's brown mustard. (Check them out on your next trip to a grocery store.)

Adjusting color using hue, saturation, and brightness controls is a lot easier than trying to adjust the amount of red, green, and blue, or the amount of cyan, yellow, and magenta, because it's close to the way people perceive colors. You'll see colors as a little muddy, for example, and know that you need to increase brightness, or you'll see grass as a little too yellow, and know that you need to move the hue setting toward green. But if you have to move sliders to make the changes, it's still not all that easy to adjust colors this way.

The problem with sliders is that first you have to figure out which sliders to move and in what direction; then you have to guess how far to move each one; and then you get to see how far off you are, so you can do it again.

It takes a fair amount of practice before you get a feel for how to adjust colors this way. (Although it takes far less practice with hue, saturation, and brightness settings than with red, green, and blue settings.) If you find that you often have to adjust color, and you care about getting it right, we'd suggest finding a program that lets you adjust colors by choosing from an array of samples, as in the Adobe Photoshop Variations dialog box shown in Figure 9-3. That way, you get to see the colors you're picking before you choose them.

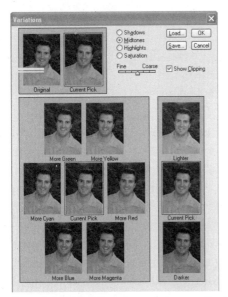

Figure 9-3 Some color adjustment features, like this one in Adobe Photoshop, show how the photo will change with each choice.

The color variations in the dialog box are all in the block of seven photos. The version showing the current color settings, identified as Current Pick, is in the middle of the block, with the six available variations arranged around it. Although the monochrome image in this book doesn't show you the actual colors in the figure, you can see the labels under each sample marking it as More Green, More Yellow, and so on.

Seeing the color shift you'll get from each available choice makes it easy to compare the choices and see which one will bring the image closer to the way you want it to look. Even better, when you choose one of the variations on color, the program moves that variation to the Current Pick position and gives you choices for additional changes. That means you can, for example, choose More Green, More Green again, and then More Cyan to get to the final setting.

This sort of global preview of what each change will do for your picture makes it far easier to adjust color. Note too the option to use Fine and Coarse changes, an option that usually goes hand in hand with this approach to color adjustment. You can set the jumps in color to Coarse to make big corrections. Then, as you get closer to the final destination, you can reset it to Fine to finish off with small, relatively subtle changes.

Adjusting for Your Printer

One of the most common reasons for needing to adjust color is that the colors don't look right when you print, even though they look fine on screen. This is most likely to happen with older printers. Recent models almost all use the same, now nearly standard, color management technologies to make it easier to translate colors from a camera to a printer. The older the printer, the less likely it is to use the same color management schemes.

Alas, a mismatch in colors between the screen and printer means you have to change the color so it looks *wrong* on screen for it to look right at the printer. Making the colors look wrong on screen is easy. The challenge is making them look wrong in just the right way.

The trick is to compensate for the printer. If you were a golfer who had a bad slice (which, for right-handed golfers, we're told, makes the ball go to the right of where you aim), you could compensate by aiming to the left of where you want the ball to go. (You could also learn how to fix your swing, but that would ruin the analogy.) In much the same way, if your combination of camera and printer tends to shift the colors toward green when you print, you can shift them in the opposite direction (which happens to be magenta) before you send the photo to the printer.

If you need to compensate for the printer this way, having a program with the kind of color adjustment feature shown in Figure 9-3 is particularly helpful. The arrangement of the variations isn't random. Take a look, for example, at the three variations in a straight line across. Figure 9-4 shows just those three.

More Cyan   Current Pick   More Red

Figure 9-4 The color variations are arranged with equal but opposite variations on either side of the current pick.

The version that's the Current Pick is in the center, with the version with More Cyan on the left and the version with More Red on the right. They are arranged this way because cyan and red are complementary colors. If you start out with a picture that looks like the More Cyan version, you can turn it into a match for the Current Pick by adding red. If you start with a picture that looks like the More Red version, you can turn it into a match for the Current Pick by adding cyan. The same relationship applies to the variations on either side of the Current Pick diagonally as well: green is the complement to magenta, and yellow is the complement to blue.

Once you realize this, it's easy to find the right correction for your printer if your editing program has a Variations screen like this. Print a photo once without color correction, and then use the Variations screen to find the closest match to the printed photo. Whatever that match is, change the color of the photo on screen by the same amount, but in the opposite direction—meaning toward the complementary color. For example, if you have to choose More Cyan twice to match the screen to the printed version, start with the original and choose More Red twice before printing. The printed version should then be a close match to the original as it shows on screen.

If your photo editor doesn't offer a visual guide to variations this way, finding the right correction will be harder, and will likely involve some trial and error, but the principle is the same. If you can figure out the color the printer is shifting the color to—in terms of red, green, blue, cyan, yellow, and magenta—you can correct for it by adding more of the complementary color to the photo. The bad news is that it takes most people a lot of practice before they can look at a photo with shifted colors and tell which of these six colors the image is shifted to. Here again, if you need to adjust colors, we'd urge you to get a program that will show you color variations on screen. In the meantime, however, Table 9-1 will help you find the complementary colors.

Table 9-1 **Complementary Colors**

Color Is Shifted to	Add Complement	Or Equal Parts of
Red	Cyan	Green and blue
Green	Magenta	Red and blue
Blue	Yellow	Green and red
Cyan	Red	Yellow and magenta
Yellow	Blue	Cyan and magenta
Magenta	Green	Yellow and cyan

Adjusting Brightness and Contrast

The auto-exposure feature in digital cameras does its job well enough that for many photos there's no need at all to adjust brightness and contrast. In some cases, however, the photo can benefit from tweaking these settings a bit. In still others, you may find that part of the photo—but only part of it—is too dark or too light. Or the problem may be that the brightness and contrast is better than you want it to be. For example, you may want to hide a face in shadow for artistic effect, instead of being able to make out whose face it is. Most photo editors have tools to let you deal with all of these possibilities and more.

Automatic Adjustments

It's worth checking out your photo editor's automatic brightness and contrast adjustment feature, which may be hiding under a name like Quick Fix, Instant Fix, Auto Fix, Auto Balance, Auto Enhance, Auto Levels, or even Auto Brightness and Contrast. Assuming your photo editor has a feature like this, you ought to try it on each photo you're considering printing or showing. Even if you don't see anything wrong with the photo as is, you'll often find that the automatic adjustment perks up the photo, as in Figure 9-5.

The original version of the photo is on the left in the figure, and the brightness and contrast are good enough so you might not think to change anything. But (in the versions that we're looking at, at least) it looks dull compared to the version on the right, which is how it looks after using the Digital Image Pro Contrast Auto Fix and Levels Auto Fix features. (We're hedging about how the two images will compare by the time they get printed, because there are all sorts of variables involved that can affect how they look. That hedge applies to just about everything we point out in this section, but we won't bring it up again.)

Figure 9-5 A single photo, before and after using the auto fix features in Digital Image Pro.

Occasionally, an automatic fix feature will make the photo look worse, but that's not a problem; simply undo the change or skip the step that actually applies it to the picture, and go back to the photo the way it was. In any case, once you try the automatic fix feature, if you're happy with the brightness, contrast, and overall detail, there's no reason to go any further. Save the file, print it, or both, and move on. If there are still some issues with brightness and contrast that you need to fix, however, there are plenty of other things you can do manually.

Manual Adjustments

Any photo editor worth the name will have brightness and contrast adjustments. In addition, most will have other settings that affect brightness and contrast in more sophisticated (read: more complex) ways. Typical controls for these other adjustments go under the names gamma; curves; tone; equalization; and dark, midtone, and bright levels (also called shadow, midtone, and highlight levels).

The details of what each of these settings does and how to set them isn't important for our purposes (especially given that the same name can refer to different features, depending on the program). What matters is that all of these features have two things in common: they affect both brightness and contrast, and they affect contrast differently at different levels of brightness. You'd probably like us to explain what we mean by that.

The short answer is that these settings can let you change the contrast in, say, the dark areas of the photo without affecting the contrast in light and midtone areas. Or they can let you change the contrast in light areas without affecting the contrast in dark and midtone areas. You'll find a more detailed answer in the section "Contrast, Brightness, and Mapping," later in this chapter. The next few photos, starting with Figure 9-6, show what this idea means in practice.

Figure 9-6 The lower half of this photo shows almost no detail.

You've seen this photo before, but you may not recognize it. We used it in Chapter 4, "Is That a Snapshot in Your Camera, or Did You Take a Photograph?" to show how putting a horizon across the center of the screen can be a mistake. When we used it in Chapter 4, however, you could see detail in the bottom half of the picture. In this version, the bottom half is a nearly solid mass of black.

You can't fix this sort of problem very well with brightness and contrast controls. You can raise the brightness to lighten up the bottom half of the photo, but that raises the brightness in the sky too, which begins to lose details in the clouds. Raise the contrast as well, to make the details in the ground stand out better, and the clouds disappear into the sky, as shown in Figure 9-7.

If your program has controls that let you change the contrast in the dark area without affecting the light area, you can adjust the picture to look like the one in Figure 9-8. As you can see in the figure, this lets you adjust the dark area so you can see detail in the ground without affecting the details in the sky at all.

Figure 9-7 Adjusting brightness and contrast brings out detail in the ground, but at the cost of losing all detail in the sky.

Figure 9-8 Your program may let you adjust the dark area without changing the light area.

We created the particular version of the photo in Figure 9-8 in Digital Image Pro by using the Adjust Lighting option and adjusting the Add Flash slider. We mention this particular approach because of all the ways to get this same result, the idea of choosing an Add Flash feature to brighten a dark area is the most intuitive approach we've seen. Unfortunately, few programs offer anything this easy.

We also managed to produce essentially this same fix in various programs by adjusting the midtone settings (for programs that let you set shadows, mid-tones, and highlights separately), changing gamma settings (gamma is a common option that changes contrast differently for different levels of brightness), and adjusting a curve that defines how shades of gray in the file look on the screen.

There are, in fact, so many different ways to do this, and they vary so much from one program to the next, that it's pointless for us to go through the details of each approach here. Instead, we suggest you look through your program for any of the features we've mentioned, or any other feature that seems to let you change the mapping for shades of gray, as discussed in the sidebar "Contrast, Brightness, and Mapping." Then check the program's manual or help file to find out how to use the particular feature. Be forewarned that in many programs the feature will be both complicated and technical. But you should be able to figure it out with a combination of the instructions that come with the program and a little hands-on experimentation. It may also help to keep firmly in mind that what you're trying to do is adjust contrast in dark areas separately from adjusting contrast in bright areas. The next section, "Contrast, Brightness, and Mapping," should help you understand the process as well.

Contrast, Brightness, and Mapping

Before you read this, be aware that we're about to talk about one of the more complicated subjects in this book, so don't be surprised if you have to go over this section a couple of times to get it. And if you skipped over Chapter 1, "Everything's Coming up Digital," it would be a good idea to go back and read it before you go any further here. You don't have to read this section to under-stand anything else in this book, which means you could choose to skip the sec-tion entirely. But the payoff if you work your way through it is that you'll have a much better understanding of how to adjust contrast and brightness in your photos. So arm yourself with a bottle of aspirin to ward off the headaches, and take a shot at it. You have been warned.

Let's start with mapping, which determines how each level of gray in the image gets displayed on screen or printed. As we discussed in Chapter 1, cameras and monitors create all the colors you can see from just three primary colors—red, green, and blue. And they need only 256 shades of each color to produce enough colors for photographic quality color. By convention, these 256 shades of each primary color are often called shades of gray.

When your camera takes a picture, it assigns a specific shade of gray to each primary color in each pixel. Call them shades, or levels, 1 through 256. When your monitor shows an image, it likewise uses a specific shade of gray for each primary color in each pixel. You might assume that if the camera assigns, say, a level 1 shade to a particular pixel, your monitor will use level 1; if the camera assigns level 2, your monitor will use level 2; and so on. This isn't necessarily true.

When you crank up the brightness on the monitor, what you're really doing is telling the monitor to assign all the shades of gray in the image to higher numbers on the monitor. So what was level 1 is now mapped to, say, level 51, level 2 is mapped level 52, and so on.

When you do this in software and apply the change to the file, the changes become permanent because they're stored in the file itself. Note too that the levels only go as high as 256. So if you crank up all the levels by 50, all the levels from 206 to 256 get converted to 256. That's why you lose details in bright areas of your photo as you move to higher levels of brightness: all the bright pixels turn into the same shade of gray. Similarly, you can lose details in dark areas if you adjust brightness downward.

When you adjust contrast, you change mapping in a different way. Crank up contrast, and you essentially put more distance between shades of gray as they existed in the original file. For example, levels 100, 101, and 102 in the original file may map to levels 100, 105, and 110. This brings out details, because it's easier to see the differences between nearby areas if the shades of gray are further apart. The price you pay is that you lose both the brightest and darkest shades of gray in the original file, since there's nothing below 1 or above 256 for them to map to.

Some photos have both light and dark areas, like a tree line against a bright sky. In many of those photos, one of the two extremes—light or dark—may have good contrast and brightness even though the other doesn't. In those cases, you need to adjust the contrast—which means changing the mapping—for one extreme without changing the mapping for the other. In short, you need to adjust the contrast differently in the dark areas and light areas, which is what we mean by adjusting the contrast differently at different levels of brightness.

One other thing: if you draw a graph with the 256 shades of gray for the original file on the x-axis (the horizontal) and the 256 shades of gray for the adjusted file on the y-axis (the vertical), you'll start out, before making any changes, with the graph as a straight line, starting at the 0 point, as in the left side of Figure 9-9.

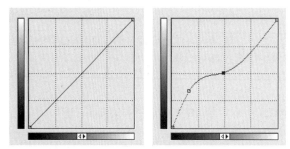

Figure 9-9 These two graphs show how the grays you start with map to the grays you end up with before making adjustments (on the left) and after (on the right).

If you adjust brightness, the graph remains as a straight line parallel to the original line, but moves the starting point to some higher level on the x or y axis. If you adjust the contrast, the slope of the line—meaning the angle relative to the x- or y-axis—changes, but it's still a straight line. However, once you start mapping the shades so different areas of brightness have different levels of contrast, the shape of the line changes from straight to curved, as in the right side of the figure. In this particular case, the mapping for the bright shades is virtually untouched, but the midtones and dark areas are changed significantly.

Some programs will actually show you the graph, and let you adjust the mapping by adjusting the shape of the curve, using the mouse to drag it to a new shape. This is the most powerful, and most difficult to master, tool for fixing problems with contrast and brightness. If your program has this feature, it's worth spending the time learning how to use it. As you drag the line, try to visualize what you're doing in terms of changing the dark pixels to lighter shades, or lighter pixels to darker shades, as appropriate. And try to visualize how that change will affect the image. With enough practice, you'll be able to predict what changes you need in the shape of the curve to adjust the brightness and contrast the way you want it.

Spot Fixes

There's one other tool—or pair of tools—for adjusting brightness and contrast that you should be aware of: burn and dodge tools. The more your photo editor qualifies as a low-end product, the less likely it is to have these tools. If it has them, however, you'll find that they can be incredibly useful.

The terms *burn* and *dodge* both come from techniques photographers use with chemical film in a darkroom. To create a print from a negative, you shine light through the negative to fall on the light-sensitive paper that will become the print. The more light that hits any given spot, the darker the spot will be. Darkroom photographers talk about *burning in* the image. If they want to darken a particular area of the photograph, they will let light shine on that area, while shielding the rest of the photo. That effectively adds extra light to the area, burning in and darkening the image just in that area. If there's an area that's too dark, however, they can lighten it in the print by shielding that area somewhat so it gets less light, a technique called *dodging*.

Lingo   *Burn* tools let you darken an image in a specific area. *Dodge* tools let you lighten it.

Photo editing programs don't deal with light, but some offer tools that will give you the same effect. If your photo editor has a burn tool, you can run it over an area to darken it. If it offers a dodge tool, you can run it over an area to lighten it.

Burn and dodge tools are best used on relatively small areas, but they can work wonders. Suppose, for example, that you're trying to sell a stained glass window on eBay, and want to post a picture of the window. You don't have a good way to show the window off to its best advantage (if you did, you'd probably be keeping it) so you lean it up against a sliding glass door and wind up with a picture like the one in Figure 9-10.

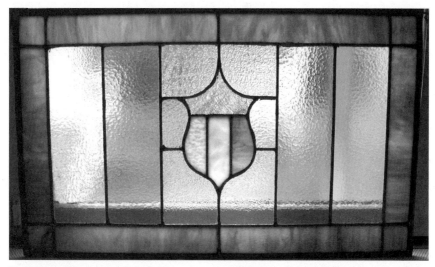

Figure 9-10 This shot is a little dark in some areas, particularly the lower left corner.

In the color version of this photo, the entire outside frame of colored glass looks dull; there isn't much light getting through it, with the door frame in the way. Even in the black-and-white version you can see that the lower left corner is far too dark. With a dodge tool, you can easily lighten the lower left corner and the outer frame. Simply turn on the dodge tool, then click and drag your mouse cursor over the area you want to lighten. Figure 9-11 shows the finished result after lightening with the dodge tool in Photoshop.

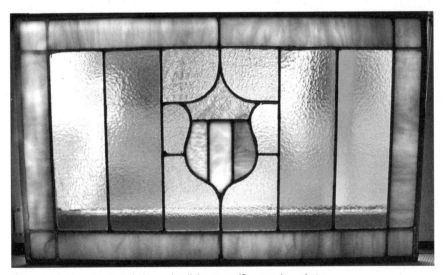

Figure 9-11 You can use a dodge tool to lighten specific areas in a photo.

You can't see the full effect of the improvement in black and white, but you can certainly see that Figure 9-11 is brighter than Figure 9-10—and only around the outside. In particular, the lower left corner is no longer lost in the dark.

Burn tools work the same way, but to darken rather than lighten. Some programs even let you use the same tool for both functions—using the mouse by itself for lightening, say, and the mouse plus the Alt key for darkening. In this book, we've used a burn tool once or twice to add shadow to faces, and make them unrecognizable, in cases when we didn't have permission to use an individual's picture.

Fixing Specific Areas in a Photo

As we mentioned, the burn and dodge tools work best for small areas. Try them on too large an area and you may leave visible streaks from successive strokes. However there are ways to fix larger areas—not just for brightness and contrast, but for most adjustments you can make.

The secret is to select what you want to adjust, and then give a command to make an adjustment. With most programs, the change—whether brightness, contrast, color correction, or anything else—will apply to just the part of the image that you've selected. The key to making this trick work is learning how to select exactly what you want to change. The hard part is finding a tool that will let you do that.

Most photo editors, even the most limited, have at least one or two tools for selecting specific areas of the photo. Some have much more than that. Common choices let you select a rectangle (including a square), ellipse (including a circle), and a freehand drawn section. (The last is generally a bad idea, because it's hard to draw the area neatly with a mouse.) Some programs also offer a polygon tool that lets you draw straight lines automatically from point to point to define a selection. An even more interesting choice is a magic wand tool that will let you click on a spot to automatically select all nearby pixels that match the pixel you click on.

You'll need to explore the choices in the program you have, and look for additional features, like the ability to add to the selection you've already made or delete part of the selection. Then try the technique with a picture that needs improvement. Consider, for example, Figure 9-12, which is the raw version of a photo we used in Chapter 2, "Knowing (and Choosing) Your Camera."

Figure 9-12 The area around the fireplace in this photo is too dark.

This is not the way the photo looked in Chapter 2. This raw version is far too dark in the area around the fireplace. We fixed it by selecting the area with a polygonal selection tool in Photoshop. First we zoomed in on the area we wanted to work with to make the selection easier. Then we chose the polygonal tool, and clicked on each successive point to extend the selection, working our way around the fireplace and back to the starting point. Figure 9-13 shows a zoomed-in view of the photo, with the selected area indicated by a white dashed line.

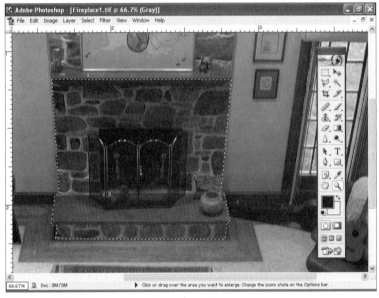

Figure 9-13 Note the white dashed line that indicates the selected area.

With the area selected, we zoomed out to see the entire photo again, and adjusted the brightness of the selection to match the rest of the photo. Figure 9-14 shows the image in midadjustment, with the brightness purposely set far too high, just to show clearly that you can adjust the selection without affecting the rest of the image.

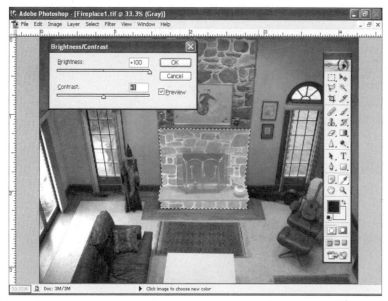

Figure 9-14 You can adjust the brightness in the selection separately from the rest of the image.

Figure 9-15, finally, shows the finished version, with the fireplace area adjusted to match the rest of the photo.

Figure 9-15 After adjusting the one area, the fireplace matches the rest of the photo.

Here again, the best advice we can give you for learning how to do this in the photo editing program you have is to first scour the program looking for selection tools, and then practice using them, looking for options that let you select exactly what you want. The better you get at selecting the areas you want, the better the adjustments to your photos will look.

Removing Unwanted Objects

Improving your photos by fixing flaws like color shifts and poor contrast is one thing. Improving them by changing the photo is a quantum jump up in concept. One of the more impressive things you can do to fix a photo is get rid of things. Whether it's a blemish on someone's face, a car in the middle of a field, or an unwanted person taking up a good portion of the shot, you really can get rid of it if you know how. Here's how.

Cleaning Up Small Areas

Let's start with something small, meaning that it's taking up just a small part of the photo. In Figure 9-16 we've used Digital Image Pro to zoom in on a close-up of a face to show just the area around the mouth and nose. If you look carefully, you can see a small mole on the subject's right check (the left side of the photo), and a small white spot—probably a flaw in the photo—just under the nose on the left side.

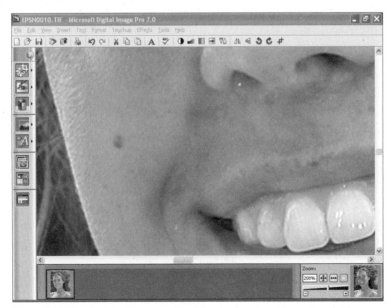

Figure 9-16 The close-up shows a mole on the cheek and a white spot under the nose.

These aren't major flaws by any means, but the white spot could be distracting if you print this at, say, 8 × 10 inches. And just for argument's sake, let's assume the woman in the photo hates the mole and would rather that it not show in the picture.

Digital Image Pro happens to have a feature called Remove Spots or Blemishes, which makes fixing things like this extraordinarily easy. Simply choose the option to get to the screen shown in Figure 9-17, move the mouse pointer over the blemish or spot you want to remove, and click. Removing the white spot took only one click. Removing the mole took four clicks, because it was a bit larger.

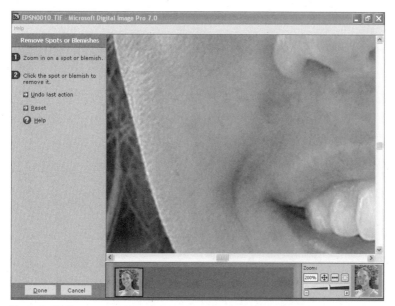

Figure 9-17 The same picture, after removing the blemishes using Digital Image Pro.

Alas, few programs offer a spot remover that's quite so easy to use, but a fair number offer a cloning tool that's almost as easy. With a cloning tool, you first click on a spot you want to clone, or copy from. To cover the mole, for example, that spot would be somewhere nearby on the cheek. Then you click on the area you want to cover—the mole in this case—and either move the mouse and click again, or click and drag the mouse back and forth to clone from the area around the first spot you chose.

The idea with cloning is to pick a spot that is similar enough to the area you want to cover so the cloned pixels will match the surrounding area. The result will be indistinguishable from the fixed version of the picture you see in Figure 9-17.

Depending on the program, you may be able to adjust the brush size and other aspects of the cloning tool as well.

Still another approach to fixing small spots like these is one we mentioned earlier—a smudge tool or blur tool. Look for a drawing tool that will let you drag the mouse over an area to smear or smudge—essentially mixing the pixels together with other nearby pixels—to effectively make the spot go away.

Removing Large Objects and People

Large objects are nearly as easy to remove from a photo as small objects, but the techniques are different. Take a look, for example, at Figure 9-18.

Figure 9-18 There's something missing from this photo, because we eliminated it.

You wouldn't know it from looking at the photo—at least, not without looking at it closely—but there's something missing. In the original, there was a highway in the frame, complete with two trucks. Figure 9-19 shows the original picture. We removed the road to give us a picture of something approaching unspoiled wilderness.

Figure 9-19 The original version of the picture, complete with the road.

To get rid of the road, we started by selecting a rectangular area in Photoshop roughly a third of the way from the bottom of the picture, copying it, and then pasting it over the area where the road curves to the left. Then we used the polygonal tool to define an odd shape from the area in the lower right corner of the picture, and copied and pasted that section. Figure 9-20 shows an early stage of the process.

Figure 9-20 Covering up an unwanted element, piece by piece.

We repeated this basic procedure with one or two additional shapes. Figure 9-21 shows the photo with the road almost entirely covered up. At this point in the process, you can see sharp lines where the copied pieces overlap.

Figure 9-21 A later stage in the process of covering up the road.

Finally, we used the cloning tool to copy small areas over the lines and hide them. Depending on the image, a smudge tool may be the preferred choice for this step. The final result is the one we showed first, in Figure 9-18.

Not all programs will give you all the tools you need to do something like this, but a surprising number do. It's certainly worth exploring the program you have to see if you can pull off tricks like this. If not, you may want to upgrade to another program.

One last tip if you try this: some programs will insert each copied piece onto a separate *layer*, the electronic equivalent of putting the copy on an overlay that sits on top of the photo. This lets you move each piece around, and even modify it without disturbing the picture underneath. If you decide you don't like what you did, you can delete the image on the layer and get the original picture back. When the time comes to blur the edges, however, be sure to merge all the layers into one. The typical command for this is to *flatten* the image.

All of which brings us to the end of the line for the subject of fixing flaws in photos. We don't want to imply that we've taught you everything there is to learn about fixing photos, but if you've followed us through this chapter and the one before, you've learned enough to fix some significant problems in your pictures. What you need to do now is practice what you've learned. Along the way, you'll pick up lots of details that you can get only from hands-on experience. If you're like us, you'll find that spending time working with your photos to improve them is so absorbing that you lose track of time when you're doing it. If so, we can only say, enjoy yourself.

Key Points

- When it comes to fixing flaws in photos, there's usually more than one way to get the same result. So if a program doesn't offer a particular tool that you're looking for, be sure to look for alternatives.

- Most photo editors offer a simple red-eye reduction feature to fix red eye. For those that don't, look for a way to paint the red pupils black.

- There are any number of ways to adjust color—and you'll often find more than one way in a given program.

- Some color adjustment features let you change the color settings by using sliders. The sliders can be used either to adjust hue, saturation, and brightness or to adjust the amount of each primary color, either using red, green, and blue as primary colors, or cyan, yellow, and magenta as primaries. Adjusting color by hue, saturation, and brightness is easier than adjusting it by changing the amount of each primary color, but no approach that uses sliders is as easy as it should be.

- The best color adjustment features show you an array of variations to pick from. This is the easiest approach to adjusting color because it lets you pick the choice that's closest to what you want, and see what you're getting before making the choice.

- If colors look acceptable on your monitor, but not on your printer, you need to first determine the direction the colors are shifted in, and then add the complementary color. If they're shifted toward green, for example, add cyan; if they're shifted toward cyan, add red; and so on.

- You can adjust brightness alone or contrast alone or adjust the two in a way that adjusts contrast differently for different levels of brightness. There are several ways to achieve the more sophisticated adjustment, and most programs have at least one of them. Find the appropriate feature in your program and learn how to use it.

■   Some programs offer dodge and burn tools that let you lighten or darken small areas by running the mouse over the area.

■   Most programs will also let you select a particular area in the photo and adjust brightness, contrast, and almost anything else in just that area. Get familiar enough with the selection tools in your program so that you can select just what you want, and you can work wonders.

■   To get rid of small distracting flaws in a picture, like a blemish on a face in a close-up, look for a spot removing tool or a clone tool. For larger areas, copy and paste from other parts of the picture. If the edges show on the piece or pieces you copied, use a smudge tool or blur tool to get rid of the edges.

Chapter 10

More Fun with Pictures: Special-Purpose Editing

In Chapter 8 and Chapter 9, we covered what you might think of as essential editing—the kind of editing that will improve your pictures, or in some cases save them from the trash bin. But there's another kind of editing too—the kind that adds special effects, text, or frames to your raw (or edited) photos and turns your photos into postcards, greeting cards, calendars, or even a screen saver for your PC. In this chapter, we'll take a look at this second kind of editing.

Most photo editors have at least some tools for this second kind of editing. In particular, almost all photo editors offer some special effects. However, some programs offer a lot more in this broader category of editing than others, and some are special-purpose programs designed just for one or more kinds of special effects editing. Programs that let you morph from one image to another are a good example.

We're going to take a different approach in this chapter than we did in the last two. Most photo editors offer similar editing tools for what we've termed essential editing. That let us give some details in Chapter 8, "Fun with Pictures: Basic Editing," and Chapter 9, "Advanced Editing: Fixing Flawed Photos," for how to solve specific problems, like getting rid of distracting elements in a photo. We can't do that in this chapter, because the range of possibilities for this second category of editing is so vast. In truth, there are so many variations that even if we had room to cover them all we'd surely to miss some.

What we can do, however, is provide you with an overview of the sorts of tools you can find, and a strategy for how to find them. Along the way, we'll discuss some specific examples. All of which should give you a good sense of the sort of thing to look for in whatever photo editor you have, as well as additional features to look for in programs you may want to consider buying.

Special Effects

Almost every photo editor offers at least a few special effects. Even better, if you take the time to explore the choices, you'll usually find that there are a lot more than you may think at first. Choices on a first-level effects menu often lead to submenus with their own list of effects, and any given effect usually has one or more variables that you can use to get very different visual results. The two photos in Figure 10-1, for example, show the same photo with the same Adobe Photoshop effect (*colored pencil*), but with different settings for the effect. Both make the photo look like a drawing, but the two images look noticeably different from each other.

Figure 10-1 Different settings for the same effect can yield different results.

Similarly, Figure 10-2 shows the same photo again with a different effect—posterization in this case. Again, different settings for the effect yield noticeably different final results. Just as important, note how different either of these images is from the first two.

Figure 10-2 Different settings for posterization can also yield dramatically different results.

The lesson here is twofold. First, the effects you choose can dramatically change the overall visual feel of the photo. Second, it pays to take time to explore the various nooks and crannies in your photo editing program. Play with the different effects available in the program you have, and with the different settings available for any given effect. Also try combining effects to see the results. You may find you've gone up a blind alley trying to do something that doesn't come out well, but you also may find something you like particularly well.

Ultimately, the only way you'll get to a point where you can predict what the settings for any given effect will do is to get some experience using those settings and seeing what they do. So go ahead and play, like a kid making mud pies. If you don't like the mud pies you make, you can cast them off and start again.

Adding Graphic Elements

In addition to options for special effects, most photo editing programs let you add text and graphics elements. At the very least, you should be able to paste graphics into the photo using the Microsoft Windows clipboard. For example, the mice in the lower left corner of Figure 10-3 are from clip art that comes with Microsoft Office. (For details about how we inserted the clip art, see the sidebar "Inserting from Office Clip Art.")

Figure 10-3 Almost any photo editor will let you insert graphics from elsewhere.

Also look for a text tool that will let you type text, format it, and move it wherever you like in the photo. Don't be shy about using large type to make sure people can read it. And be sure to use light colors against dark backgrounds, or dark colors against light backgrounds.

Inserting from Office Clip Art   Getting the graphic into the photo with the elephant was a little harder than you might expect. We first inserted it into a Microsoft Word document, leaving plenty of blank space around it. You would probably assume that you could then select the clip art image, copy it, and then paste it into your photo editor, but when we did that, the image was cut off on one side.

To work around the problem, we first held down the Alt key and pressed the Print Screen key to copy the Word window to the Windows clipboard as a graphic. Next we opened a new, blank file in Photoshop and pasted in the image from the clipboard by choosing Edit, and then Paste. Then we used the selection tool in Photoshop to select the area around the mice and copied the image to the clipboard by choosing Edit, followed by Copy. Finally, we switched to the elephant photo, which was already open, pasted the graphic into the photo—by choosing Edit, and then Paste—and moved the clip art to the position where you see it in the figure.

One other feature you'll find in most photo editors is the ability to draw lines, often with a choice of paint tools. Typically, you can draw lines with hard or soft edges using different shapes and sizes for the brush. You may also be able to spray paint or airbrush an area. Experiment with all these tools to see how they work. Alas, unless you're a pretty good artist, you're more likely to ruin your photos with most of these tools than to improve them, but don't let that stop you from experimenting with them.

Adding Frames and Cutouts

Once you get past the basics of pasting in graphics from elsewhere and inserting text and lines, all bets are off about what your program can and can't do. You need to explore each menu choice to find out what other features are available, and be on the lookout for features in unexpected places.

One feature you'll find in a fair number of programs aimed at nonprofessional photographers is the ability to add frames to a picture, and also apply cutouts. Figures 10-4 through 10-6 show some typical examples, but keep in mind that for those programs that have these features at all, the particulars vary tremendously from one program to the next. In most cases you simply pick a template and let the program insert the photo in the right place. In some cases you can then adjust the position and size as necessary.

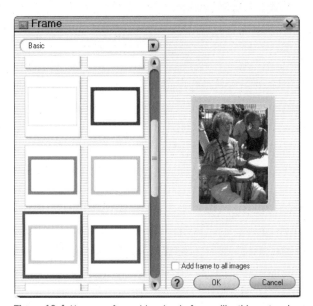

Figure 10-4 You can often add a simple frame, like this rectangle around the image, added with Olympus Camedia Master's set of basic frames.

Figure 10-5 More elaborate frames like this one in ArcSoft PhotoImpression are also available in many programs.

Figure 10-6 Also look for a cutout feature like the octagonal shape around this photo and the other shapes in the bottom third of this PhotoImpression screen.

Postcards, Greeting Cards, and Calendars

Many photo editors also let you insert your photos easily in a variety of formats—including postcards, greeting cards, and calendars. Here again, for photo editors that offer this feature, the particulars vary wildly from one program to the next. Figures 10-7 and 10-8 show two examples of the sort of capability you may want to look for.

Figure 10-7 Sometimes, as with this New Year's card template in Olympus Camedia Master, you can find a card that exactly fits a particular photo.

Figure 10-8 Calendar features, like this one in PhotoImpression, not only provide a format, but typically calculate the days for the calendar month.

From Photo to Screen Saver or Wallpaper

There are two common features in photo editing programs worth special mention: the ability to turn your photos into a screen saver and the ability to turn photos into *wallpaper*, the background for your Windows desktop.

Lingo   *Wallpaper* in Windows is a background image for the Windows desktop.

These features usually require little more than picking the photo you want to use and giving the command to turn it into wallpaper or a screen saver. With screen savers, you often get to pick multiple photos—sometimes defining them first as an album—and the screen saver turns them into a repeating slide show.

Most programs that let you turn photos into wallpaper, screen savers, or both, also have commands to remove the setting. But whether your program has that capability or not, it may be helpful to know how to change these settings manually in Windows. If you right-click anywhere on your desktop—except on a shortcut—and you then choose Properties from the shortcut menu, you'll open the Display Properties dialog box. Depending on your version of Windows, you'll see either a Desktop tab or a Background tab. Choose the tab, and you'll see a list labeled Background or Wallpaper, depending on the Windows version you're using. If you pick an item from that list and then choose OK, that item will show as wallpaper—the background to your desktop.

When your photo editor program turns a photo into wallpaper, it adds that photo to the list, and sets Windows to use it. You can change the wallpaper at any time—or get rid of it—by opening the Display Properties dialog box, choosing the Desktop or Background tab as appropriate, scrolling through the Background or Wallpaper list, and choosing a different item on the list, or choosing None.

One of the other tabs in the Display Properties dialog box is labeled Screen Saver. Choose this tab, and you'll see a Screen Saver drop-down list. As with wallpaper, when your photo editor sets a photo or group of photos as the screen saver, it adds the name for the screen saver to the list, and sets Windows to use it. Here again, you can change the screen saver at any time by opening the dialog box, choosing the Screen Saver tab, and then choosing a different item on the list, or choosing None to turn off the screen saver.

One note about wallpaper: resist the temptation to use any photo that has a lot of detail in it. You may have an oriental rug that you love, but turn a photo of it into wallpaper, and you may have trouble finding your shortcuts on your desktop, as in the wallpaper shown in Figure 10-9.

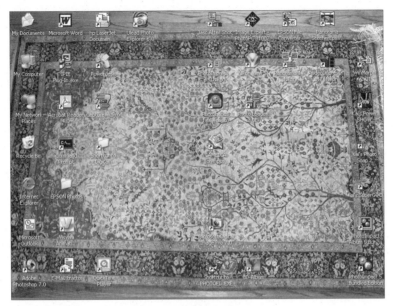

Figure 10-9 A busy photo makes questionable wallpaper.

You'll be much better off if you use a simple, uncluttered photo. Failing that, look for a template in your program that will let you insert one or more photos into a small area, but leave most of the screen uncluttered, as in Figure 10-10. We created this wallpaper from a template in Olympus Camedia Master.

Figure 10-10 Some programs provide templates for wallpaper, so you can insert your photos in wallpaper without cluttering up your screen.

Special-Purpose Editors

We can't leave the subject of special-purpose editing without at least mentioning some programs, and types of programs, whose capabilities might best be described under the category of weird editing.

First on the list is Kai's SuperGoo from Scansoft, which offers two very different capabilities. The Goo feature lets you stretch an image in various ways till it's well beyond the point of being recognizable. It also lets you stretch the same image in any number of different ways, save each version to a filmstrip, and then run an animation showing the image twisting and turning, to get from one stored frame to the next.

The Fusion module lets you join part of one image with part of another—melding the top of one person's face (hair, eyes, and nose), for example, to the bottom of another's (mouth and chin). If you've ever watched Conan O'Brien's projections of what the children of various pairs of celebrities would look like, you've seen similar photos. The result is often worth a chuckle (well, it gets laughs on O'Brien's show at least), and the process itself is fun.

Finally, you might want to try your hand at morphing. You've certainly seen the results of morphing in everything from political ads to Michael Jackson videos. There's no reason you can't get in on the fun, using your own photos. The trick for morphing faces is to start with two photos that are generally similar in composition. For morphing people, for example, you should start with two faces that are essentially the same size and facing at the same angle. Typically, you then use a drawing tool to define some number of points on the first person's face and match them to equivalent points on the second person's face. Choose a few simple options, like how many steps to use in morphing from one face to the other, and everything else is automatic.

One freeware morphing program you might want to try is Winmorph. You can find a link for downloading it at *www.debugmode.com/winmorph*. If you want to find some alternatives, go to your favorite search engine (*www.altavista.com*, *www.google.com*, or *www.yahoo.com*, for example) and search for *morphing software*.

Key Points

- Almost all photo editors offer special effects of some kind, but they vary from one program to the next.

- Most programs offer more variations for special effects than they appear to at first glance. Most effects have multiple settings, and a different combination of option settings can dramatically change the look of a picture.

- You can add graphic elements to virtually any photo editor by copying graphics from other programs and pasting them into the photo.

- Many photo editors can help you automatically insert frames around photos; use cutouts in various shapes; add photos to postcards, greeting cards, and calendars; and even turn photos into screen savers and Windows wallpaper—the background for the Windows desktop.

Part III

Sharing Your Photos

Taking pictures can be a lot of fun. So can editing them to improve the way they look or to add enhancements ranging from simple frames to morphing from one image to another. Ultimately, however, taking and editing photos is pointless if you don't have some way to look at them when you're done. And most people want to share their photos with others.

Historically, sharing your photos has usually meant having prints or slides to look at. With digital photos, it can also mean viewing them on a computer monitor or TV set, posting them on a Web site, or sending them by e-mail.

In this part of the book, we'll cover all these possibilities—printing photos, viewing them on screen, and sharing them electronically. We'll also cover a few more variations— like adding photos to a letter or newsletter for friends or clients. Along the way, we'll give some practical advice about things like how to choose the right paper for printing and how to keep files down to a reasonable size for e-mail so they won't bounce back because they're too large, and it won't take an unreasonably long time to send or receive the image over a modem. We'll finish up with a look at some Web sites of particular interest for digital photography, primarily sites that let you store and share your photos.

Chapter 11

Printing

Many of the choices for looking at your photos are no different with digital photos than with chemical film. You may want to frame your pictures, put them on your desk, hang them on the wall, mount them in an album, or keep a copy in your wallet or purse. You may also want to give copies to the people in the photo. All of these possibilities have one thing in common: you have to print the photos first.

For most people, printing remains the default choice for showing and storing their digital photos—at least the ones they really want to keep around. That makes printers an important subject for digital photography, which is why we talked about the various kinds of printers and what you can expect from each in Chapter 5, "Special Issues for Digital Photography." In this chapter we'll focus on how to take best advantage of whatever printer you have by picking the right paper, the right ink, and the right settings in the printer driver.

We'll also discuss how to print your photos without needing a printer, or even a computer. We know that printing without a printer sounds a little like the Cheshire cat who disappeared and left his smile behind, but you really can print high-quality photos even if you don't have a high-quality printer, or any printer at all.

Check Your Driver

At the beginning of this book we mentioned that when it comes to film, if there are any obvious problems with color during the developing process, it's up to the technician doing the developing to compensate for it—a comment that also applies to any other special requirements for developing, including such basics as setting the size for each print.

We also mentioned that one of the big differences with digital photography is that if you print your own pictures, you're the technician in charge. If you want several photos on a page, or one photo spread out over several pages, or you want to print a color image in black and white, with or without a sepia tinge, or do anything else beyond accepting the default settings for printing, you're the one who has to do it. If you're not feeling particularly competent in that role, there are ways to fix that.

The single most important thing you can do to improve the way your pictures look on a given printer is to get familiar with the printer driver—the software that gives the printer its marching orders. Most people know that they can use the printer driver to change the paper type or the quality setting, but few ever go much beyond that to explore the driver. That's a big mistake. Most drivers give you a fair amount of control over your output. Some give you a lot of control. If you don't know what features you have available, you might as well not have them.

Opening Your Printer Driver

The first step for getting familiar with your driver is to look through it and see what's there. In general, you can open the driver in either of two ways. You can give the print command from within a program, and then choose Properties to see the driver's Properties dialog box, or you can go to the Windows Printers And Faxes dialog box (called the Printers dialog box in some versions of Microsoft Windows), and open the driver from there. We suggest using the second route. If you open a driver from within a program, you may not see all the options.

To open the driver, choose Start on the Windows taskbar. Depending on your version of Windows and how it's set up, you may then see an option for Printers and Faxes. If so, choose it. If you don't see that option, choose Settings and then choose either Printers or Printers and Faxes, depending on your version of Windows. Any of these choices will open the Printers And Faxes dialog box or the equivalent Printers dialog box depending on the version of Windows. Find the icon or text entry for the printer driver you want to explore, right-click on it, and choose Properties.

At this point you'll have the driver open, with several tabs showing, as displayed in Figure 11-1.

Figure 11-1 Here's one example of a printer driver as it shows in Microsoft Windows XP.

The specific tabs you see in the driver will depend partly on the version of Windows and partly on the driver. Whatever tabs are available, choose each one in turn, and browse through them, looking at the available options. As you browse, be sure to choose any buttons that suggest they'll show more options—particularly if they have names like Custom or Advanced. Most of these will open dialog boxes with additional settings. In particular, in Windows XP Professional, be sure to choose the General tab, as shown in Figure 11-1 and then Printing Preferences. On most drivers, this will open a new dialog box with its own set of tabs that include the settings we're most interested in for this discussion.

What to Look for in Your Driver

Some of the options you'll find in the driver, like the choices for Sharing, Ports, and Color Management, are standard Windows features that remain the same from one printer to another. Others, however—including options for paper size, color corrections, and print quality—depend entirely on the individual printer. These can vary dramatically from one printer to the next—in the choices available, in how they are organized, and in how they look and work.

To get a sense of the differences, take a look at Figures 11-2 through 11-4, which show the screens for the printer-specific settings for three different printers—an older Tektronix (now Xerox) laser printer, and newer Epson inkjet and Xerox laser printers.

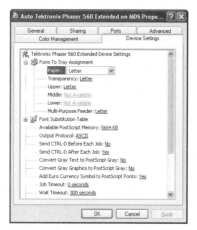

Figure 11-2 The printer-specific screens for many older printer drivers offer a forbidding list.

Figure 11-3 Newer printer drivers tend to offer fewer choices on each screen, making them less overwhelming.

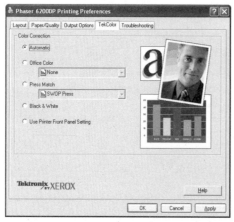

Figure 11-4 Many newer drivers also offer thumbnail images that change as you change settings, to show how each setting will affect the printed image.

Because there are so many variations on what you'll find in a printer driver, we can't possibly cover them all here. However, we can give you some pointers about the most important settings to look for and experiment with. We'll focus on the settings that are of particular importance for photos.

Paper Type or Media Type

In a later section, "Choosing the Right Paper (and Ink)," we'll discuss the various papers you can print on. Whatever paper you choose, however, it's important that your printer knows what paper you're using, since different types of paper need to have the ink laid down in different ways.

A few printers offer an automatic setting that can actually sense the type of paper you're using and adjust printing accordingly. However, most printers give you a choice of paper type, as shown in Figure 11-5.

Figure 11-5 Some printer drivers, like this Hewlett-Packard Deskjet 5550 driver, list more paper types than others.

Typical options include plain paper, high-quality coated paper, glossy photo paper, matte photo paper, transparencies, and so on. Unfortunately, the names for the paper types, and even the number of choices, vary from one printer to the next, and matching the option to the paper is sometimes a matter of guesswork. If you see two or three options that may apply to a given type of paper, try each one to see which works best.

Paper Size, Paper Source, Number of Copies, and Orientation

These are all fairly cut-and-dried settings. The Paper Size setting should match the paper you're using, and the source should match the input bin, slot, or paper roll that you're using. Set the number of copies to match what you want, and set the orientation to print in portrait (with the top of the image along the short side of the page) or landscape (with the top of the image along the long side of the page). You may also have a choice to center the photo on the page. If your printer is designed to use paper rolls—an option you'll find in some photo printers—look for a banner mode, to let you print a panorama that's much wider than it is high.

Quality Modes and Features That Affect Quality

You'll often find settings for printer resolution in dots per inch, particularly in older printer drivers. In most of these cases, the higher you set the resolution, the crisper the photo will be. However, we've seen printers that print their best quality photos at one step down from the highest resolution. We've also seen printers that show little or no improvement between the second highest and highest resolution. Given that it generally takes longer to print at higher resolutions (because there are more bits of data to deal with), be sure to experiment with different resolution settings to confirm which one gives you the best looking results, and also to see whether the better results at higher resolutions are better enough to be worth the wait.

You may also find a setting for dithering, which is a technique printers use to create most of the colors they print from just four or six ink colors. All you need to know about dithering for our purposes is that there is more than one way to do it. Some printer drivers let you choose which approach to use. Dithering choices are most likely to show up in a custom setting feature. In most cases, if you have the choice available, the driver will give you some indication of which one is best for photos. If it doesn't, you'll have to experiment to find out.

Newer printer drivers often hide the resolution settings and provide a slider or buttons instead with *Quality* at one extreme and *Speed* at the other, as in Figure 11-3. Most often in these drivers, the quality mode sets the resolution, with higher quality modes printing at higher resolutions. In some cases you can choose a custom setting that lets you set the resolution to an even higher level.

If you see quality choices like Text, Text and Graphics, and Photo, the settings are probably controlling both resolution and dithering scheme. They may control other features as well.

One setting for inkjet printers that's often hidden in vaguely defined quality settings is the choice between unidirectional and bidirectional printing. With bidirectional printing, the print head lays down ink both when it moves left to right across the page and when it moves right to left. This takes less time to print a page, but often leaves telltale bands. With unidirectional printing, the print head lays down ink only when going in one direction, which makes it easier for the printer to line up the successive sweeps and minimize banding.

Some inkjet printers can also use a microweave technique, which overlaps the successive sweeps in a way that minimizes or eliminates banding. If your printer offers these features, you may be able to control them manually to turn them on and off, or they may automatically turn on or off depending on what quality settings you choose.

Color Management, Brightness, Contrast, and More

Most recent printer models handle color so well that you should rarely need to do anything but leave the printer in automatic mode for color management. With older models you may need to adjust color fairly often, however, and even with newer models, you may need to adjust it occasionally. If there's no automatic mode, look for an sRGB setting, which is the default choice for the color management standards that most cameras, scanners, and printers use today. Figure 11-6, for example, shows an Epson driver that offers sRGB as a choice under Color Management in the upper right.

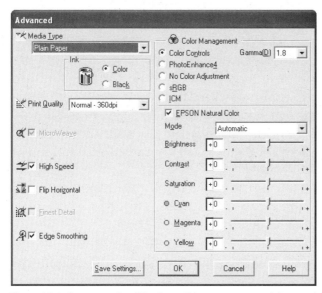

Figure 11-6 Custom settings like these give you more control than predefined quality modes.

Many printers offer a choice of color management options (sometimes labeled as color matching), often with names like Photo, Graphic, Vivid, and so on. However, the names of these options often don't match their best use. We've seen printers with a Photo color match option that print much better looking photos if you use other choices on the list. Our suggested strategy is to try the automatic setting first, if one is available. If you're not satisfied with the results, try any setting that suggests it's for photos. If you still don't like what you see, try the other settings as well.

Some drivers even let you adjust features like brightness, contrast, saturation, sharpness, or the levels of cyan, magenta, and yellow, as with the Epson driver in Figure 11-6. We'd generally rather adjust these settings in a photo editor, but if you have to consistently adjust the colors the same way for every picture you print, and particularly if the driver lets you save the settings, you may want to make the adjustment in the printer driver. You may be able to adjust the colors once, save the adjustment to disk, and no longer need to adjust the colors on all your photos. It's certainly worth a try.

Other Features

Also be on the lookout for other settings and features that can affect your photos. A clogged nozzle, for example, can leave streaks on the page indicating an absence of one color of ink. Most inkjet printers offer utilities for printing a pattern that will show you whether a nozzle is clogged, along with utilities for cleaning clogged nozzles. Other utilities will let you align the print heads to make sure vertical and horizontal lines will be straight. Some printers automatically run their alignment routine whenever you change a cartridge. With others, you have to remember to run the alignment manually.

You may also find other useful features, like the Poster option in Figure 11-7, which lets you print a single-page image—including a single photo—over multiple pages. The choices in the figure are for 4, 9, and 16 pages. Note that you may find a similar feature in your photo editing program as well. If so, be careful not to use both poster features at once; they may interact with each other in odd ways.

Figure 11-7 Some printer drivers offer a Poster feature to let you enlarge a photo to print on several pages.

Figure 11-8 is a photo printed on four pages with the poster feature. We've separated the pages a bit to make them stand out as separate pages. Note also that we've trimmed the borders in a way that will let us overlap adjoining edges and paste the pieces together, without any borders showing between the pieces.

Figure 11-8 Poster features let you print large-sized images like this one.

As you look through your driver, keep in mind that the point of this exercise is to become familiar with the available features. Be sure to experiment with the ones that look interesting, or that you don't understand fully. It's the best way to learn how they affect your output.

Choosing the Right Paper (and Ink)

When it comes to making sure your photos print so they'll look the way you want them to look, knowing about paper and ink is a close second in importance to knowing your printer driver. The good news is that your basic choices are straightforward.

Paper and Ink Basics

Paper for inkjet printers comes in four basic types: plain paper, high-quality paper, coated paper, and photo paper. (There are some variations, like greeting cards and transparencies, but we'll stay with the basics.)

Plain paper most often translates to the 20-pound weight copier paper that you probably use for most of your printing. It typically costs a penny a page or less. Many printers today can print reasonably good quality photos on plain paper, but the output falls far short of true *photo quality*—which we define as indistinguishable from a film print. With plain paper, colors tend to be dull, fine details tend to disappear, and you lose the subtle shading that, for example, gives a close-up of an apple a sense of three-dimensional roundness.

Lingo   *Photo quality* or *true photo quality* indicates a computer-printed photo that's indistinguishable from a print using film.

High-quality paper is basically a brighter white version of plain paper. In some cases it's even called bright white. The price is roughly 2 cents per page. Photo quality is a bit better on high-quality paper than on plain paper, but still noticeably short of true photo quality.

Coated paper has a different feel from plain paper and high-quality paper. In some cases the coating is extremely obvious. In others, it's subtle, so the paper feels a bit heavier than plain paper, but not all that different. The price is typically 10 cents per page, give or take a few pennies. As the name implies, coated paper has a coating designed to keep ink from being absorbed into the paper. That minimizes the loss of sharpness that comes from the ink spreading out as it's absorbed. It also makes the colors a bit punchier, since the inks are sitting closer to the surface, or on it. Photos on coated paper can approach true photo quality, but colors still look dull, and subtle shading gets lost compared to the real thing.

Photo paper, finally, is designed specifically for printing photos. It imitates the look and feel of film prints, with the goal of providing output that's indistinguishable from film prints. Photo paper comes in all sorts of variations, including glossy, semigloss, and matte, to match the variations for film prints. Typical prices run from about 50 cents to $1 per page. Given a good enough printer, photos on photo paper can be true photo quality. Most current inkjet printers can at least come close to that goal, with near photo quality.

We mentioned the price for each type of paper because that should be part of the equation when you choose paper for a particular print job. If you're printing a photo of a friend's wedding so you can frame it and give it to the couple to put on their living room wall, you probably don't care if you're spending a dollar for the paper. If you're printing the same photo to post on the office bulletin board with a thumbtack in it, plain paper at a penny a page will probably do.

Print on the Right Side

One last basic bit of information about paper is that most paper has a right side to print on and a wrong side. This is obvious for glossy photo paper, which is generally very different on one side than the other, but it's just as true for most other papers, including plain paper. When you take the paper out of its box or wrapping, look for some indication of the right side to print on. Often, you'll find an arrow with the arrowhead on the package pointing in the same direction as the printing side of the paper in the package. In a few cases, you'll find a specific statement that you can print on either side.

If the paper is already out of the box, you can sometimes spot the right side to print on because it's noticeably whiter than the wrong side. The printing side also tends to be smoother, which you can sometimes tell by brushing the paper against your lip.

Ink Basics

The basics for ink are even easier than for paper. Your printer either has special photo ink or it doesn't. If it does, the photo ink may be in a photo color cartridge that swaps out with the standard color cartridge, or it may be in a cartridge that replaces the black ink cartridge, typically providing black plus light cyan and light magenta. Check your printer manual to see if the printer offers photo ink cartridges. If it does, be sure to use them when you want the best quality for your photos.

Beyond Basics

Once you get past the basics, there's plenty of room to experiment. For example, most printer manufacturers sell their own brand of inkjet papers, often selling more than one variation on any given paper type. You'll certainly want to experiment with at least the variations available from the printer manufacturer. You may also want to experiment with paper from third-party manufacturers, a term that indicates they are neither you (the first party) nor the printer manufacturer (the second party). Kodak, for example, sells papers that are meant for any brand of printer. Keep in mind, however, that ink and paper interact with each other, so that colors may look different on different brands of paper.

Inks from third-party manufacturers are more problematic than paper from third-party manufactures. As we've suggested elsewhere in this book, we're skeptical of inks from sources other than the printer manufacturer—at least for photos.

When printer drivers choose how many dots of each color ink to use to create a given color in a photo, they start by assuming specific colors for each ink. If those primary colors are different from what the drivers are designed for, the colors in the image will be off. Keep in mind too that each printer manufacturer uses different inks—and in many cases manufacturers use different inks for different printer models. The inescapable conclusion is that a single set of inks from an ink manufacturer can't possibly match the inks in every printer, or produce the right colors on every printer for printing photos.

Some ink manufacturers sell a different set of inks for each printer model. In theory, they could match the ink colors to each printer manufacturer's inks, but, as we've mentioned elsewhere, it's not easy to match colors under all lighting conditions. We're not aware of any independent lab that has ever done sufficient testing to prove whether any ink manufacturers' inks match the original under all conditions. Until someone does, you may want to experiment with the inks yourself, or you may decide to just stay with the printer manufacturer's ink.

No Computer? No Printer? No Problem

If you've read this book from the beginning straight through, you already know much of what we have to say in this section, but we thought it would be useful to have it all in one place.

First, you don't need a computer to print your digital photos. As we pointed out in Chapter 6, "Keep Those Pictures Coming: Batteries and Digital Film," one of the ways manufacturers justify calling a printer a photo printer is by including slots that you can plug your digital film into to print without need of a computer. Some also let you connect a camera directly to the printer by cable to print without using a computer.

Printers that offer this feature generally include a built-in liquid crystal display (LCD) and menu system that lets you see your photos, print an index (a set of thumbnail images of the photos on the memory card), choose which pictures to print, specify how many to print, choose what size to print them, and then give the command to print.

If you have a computer, but not a printer—or at least not one that's good enough for printing photos—that's not a problem either. You can send your photos to an online service to have them printed and sent back to you. Figure 11-9, for example, shows the PhotoAccess Web site at *www.photoaccess.com*. As you can see in big text in the figure, the site even offers 10 free prints (at the time we wrote this at least) to encourage you to get started.

Figure 11-9 Some Web sites let you order prints made from your digital photos.

Other similar sites at this writing include *www.ofoto.com*, *www.shutterfly.com*, and *www.snapfish.com*. All of these, as well as still other sites, let you upload your photos, order prints, and then sit back and wait for them to be delivered to your (real) mailbox. In most cases, you can also edit the images online to crop them, remove red eye, and so on. We suggest exploring all four sites we mentioned. For a still longer list, go to your favorite search engine (*www.altavista.com*, *www.google.com*, or *www.yahoo.com*, for example) and search for "print photos online" (without the quotation marks), and then look through the list of hits for other sites.

Another way to print if you don't have printer (or even a computer) is to find a nearby store with a kiosk—or the equivalent, like the Aladdin Picture Center shown in Figure 11-10—for printing photos from digital film. (We'll refer to any variation on this concept as a kiosk from here on.) We mentioned these kiosks in Chapter 6, "Keep Those Pictures Coming: Batteries and Digital Film," when we talked about the new generation Kodak kiosk that will write your files to a CD-R disc. You can find this sort of kiosk in drug stores, supermarkets, photo retail stores, and elsewhere. You may even have walked by one in a local store without noticing it. Now that you know they exist, keep an eye out for them.

Figure 11-10 The Aladdin Picture Center shown here isn't exactly a kiosk, but it serves the same purpose.

With any kiosk you come across, the procedure for printing photos should be similar to what we described for photo printers with built-in slots, except that the LCD is larger. Here again, you choose which picture to print, how large to print it, and so on. In most cases, you can also crop a photo before printing, and you may be able to adjust things like brightness, contrast, and color. You may even find choices to turn your photos into greeting cards, calendars, and the like.

Because these kiosks are designed for anyone to use, they're usually self-explanatory and easy to work through, but they also have the advantage of having someone in the store who can help you if you're having a problem figuring out how to use them.

How to Print Your Photos in Wallet Size, Life-Size Blow Up, or Anything in Between

There are two issues involved for printing at any given size. The first is knowing what resolution to use for that size. The second is making sure you have a way to print at the size you want.

Throughout this book, we've talked about printing photos at 150 to 200 pixels per inch (ppi) as the rule of thumb for getting good quality output. What we haven't mentioned is that this rule applies only for images you expect to view from normal reading distances.

The next time you're standing in a train station, take a close look at the posters. Stand about a foot away from a poster with a photo, and you'll see all sorts of details from the printing process, often in the form of large, easily visible dots. Move back a bit, and the dots disappear, merging into the photo. The point is that the poster is designed to be viewed from a distance, which means the original image doesn't need a high enough pixel resolution to give you a high-quality image from just a foot away.

The same logic applies to any large images you plan to print. There's no need to use the same resolution in ppi for the original photo, as snapped by your camera, that you would need for something that you expect people to view from just a foot or two away. A good rule of thumb, subject to modification on a case-by-case basis, is to start with enough pixels in the original photo to give you the quality you demand at 8 × 10 inches. Then resample the image before printing to give you 150 ppi at the size you plan to print.

Where to Find a Big Enough Printer

Most desktop printers print on anything up to and including legal size paper. If you want to print anything larger than that, you need to go beyond the usual. One choice is to use the poster feature that we mentioned when talking about printer drivers. Expand the image to fill 16 sheets of 8.5 × 11-inch paper, arranged four up and four across, and you can print a photo at 34 × 44 inches. As we suggested earlier, if your printer driver doesn't offer this feature, you may find it in your photo editing program.

The drawback to using a poster feature is that after you print, you have to paste the pieces together. And since most printers have nonprintable areas, you'll probably have margins along each page that you have to trim off first. Much better is to use a printer that can handle larger page sizes, although that may be easier said than done.

It's not hard to find tabloid-size color printers, using either laser or inkjet technology. Many offices have them, and so do graphic artists. In fact, most of these printers can handle paper an inch or so larger than tabloid size in each direction. If you print anything at this larger size very often, you may find it worthwhile to buy a tabloid-size printer yourself.

To print anything larger than tabloid size, you need a *wide format printer*, also known as a *large format printer*. Large format printers can print on sheets of paper, but they mostly print using rolls, which means they can print as wide as the roll, and as long as necessary. Depending on the printer, they may be able to handle rolls as large as 72 inches wide.

Most wide format printers spend their lives in service bureaus and online services that will print files at poster size for a substantial price. For example, one online site we checked charges about $50 for a 30 × 40-inch print and almost $200 for a 60 × 77-inch print. At that price you won't want to print posters very often, but at least you know it's possible for special occasions.

To find a list of Web sites that print poster-sized photos, go to your favorite search engine (*www.altavista.com*, *www.google.com*, or *www.yahoo.com*, for example) and search for "print large format photos" (without the quotation marks), then look through the list. One of the more interesting sites we turned up this way was *http://mycustomprints.com*, which can calculate a recommended viewing distance for any given photo you print at any given size.

To find a recommended viewing distance go to the *mycustomprints.com* home page, and choose Image Eval Wizard. Then enter the file name of the file as it exists on your disk (you can choose Browse to navigate to the file the same way you would in a Windows program). Also enter the image type (digital camera), and either the width or height you want to use for printing the image. Then choose Evaluate File, and the site will calculate a price and a recommended viewing distance. Leave the size blank, and it will calculate a recommended size.

Note that when we tested the feature, it wouldn't read the resolution of a TIF file properly. To get the calculation working, we had to open the file in a photo editor and save it as a BMP file first. One other issue that isn't immediately obvious is that once you've calculated the viewing distance once, you can change the size information and recalculate to see the difference the different sizes make. Be sure to enter only the width or only the height and let the wizard calculate the other dimension.

Other sites we found in our search included *www.jumbogiant.com*, and *www.simivalleyphotolabs.com/printing.htm*. You might want to take a look at these as a starting point if you need large format printing, but be sure to check out things like the warranties the sites give on picture quality. The price for large format printing is a bit high to find out that you're left holding the bag if the image doesn't look good.

Key Points

- Most printer drivers offer a fair amount of control over the output. It's important to get familiar with the driver for your printer to see what features are available that can affect your photo quality.

- Learn how to specify the paper you're using, and also find out which choice to make for the paper you want to use.

■ If your printer can use paper rolls, look for a banner option that will let you print a panorama or similar image that's much longer in one direction than the other.

■ Be sure to explore the quality settings. The options vary widely from one printer to the next, and the newer the printer the more likely the driver is to hide details like resolution and dithering settings. Try to find out what the quality settings for your printer do, and look for a custom mode that gives you more control over some or all settings.

■ Some printer drivers let you adjust brightness, contrast, and even color. This is usually best done in a photo editing program. However, if you find that you consistently have to make the same adjustments, try making them in the printer driver and saving the changes.

■ Also look for maintenance features, like tests for clogged nozzles and utilities for unclogging them.

■ Choosing the right paper is also critical. There are basically four kinds of paper with variations of each: plain paper, high-quality paper, coated paper, and photo paper. Each step up the scale will give you better quality photos, but each step also costs more per page. Be sure to pick the right compromise between price and quality for the print job at hand.

■ Most papers, including plain copier paper, have a right side and wrong side for printing. Look at the packaging to find out which side to print on, and load the paper into the printer accordingly.

■ Make sure you know whether your printer has special ink for photos. If it does, make sure you use it when it's important to get the highest photo quality.

■ Some printers let you plug the camera or camera memory into the printer and print directly from the camera's memory card, so you don't need a computer.

■ If you don't have a high-quality printer, consider sending your files to an online site that will print your photos and send them back to you.

■ Another alternative is to print at a stand-alone kiosk, which you can often find in locations ranging from drug stores to supermarkets to photo stores.

■ You can also find Web sites that will print your photos at poster size, although they tend to be expensive.

Chapter 12

Viewing Photos on Screen

We don't know how many people, if any, do this anymore, but there was a time when people would go on vacation and take a whole bunch of slides—we mean slides on film, the kind that you have to put in a slide projector and project on a screen. Then they'd invite their friends over for an evening to bore...uh, we mean, regale them with their pictures and stories about their trip.

We don't have any way to prove it, but we suspect that videotape has largely taken over the role that slides used to serve in this social ritual (with digital video coming on strong). The advantage of video is not so much that it offers moving pictures instead of stills, but that it's a lot easier to put a tape in a VCR than it is to load a carousel with slides, set up a screen and projector, adjust focus, and turn down the lights so you can put on a slide show.

Digital photography levels the playing field. No longer do you need a screen, a projector, or low lighting. Most cameras aimed at consumers rather than professionals come with photo editing programs that include a slide show feature so you can create slide shows to see on your computer screen. If you prefer, you can use a full-function presentation program for more sophisticated slide shows. You can also easily record the output from your computer on a videotape so you can see the slide show on your TV, and with some cameras, you can simply plug the camera into your VCR to record the photos or pass them through to your TV to show them on screen. You can even move your photos to your Pocket PC to give a handheld slide show, or just keep more photos with you than will fit in a wallet.

In truth, there are more variations on this theme than you can shake a proverbial stick at. We won't try to cover all of them in this chapter, but we will cover the ones we consider most important, and mention a few others in passing.

Creating a Slide Show to View on a Computer

If you check your photo editing program, you may find that it has a built-in slide show feature. Typically this involves either creating an album within the program and inserting the photos for the slide show into the album, or simply creating a folder on your disk and moving the photos to the folder. To see the slide show, you first open the particular album or move to the particular folder from within the program. Then you give the command to show the slide show.

Figure 12-1, for example, shows Jasc After Shot, which can turn all the photos in a folder into a slide show.

Figure 12-1 To create a slide show in many programs, you first move the slides you want to include into a single directory.

As you can see in the figure, the current folder is highlighted on the left, and thumbnails of the photos in that folder show in the middle and right side of the screen. Also note the Slide Show icon just under the File and Edit commands. To start the slide show in this particular program, you click on the Slide Show icon or choose Tools and then Slide Show. Figure 12-2 shows the actual slide show.

EPSN0006.JPG

Figure 12-2 Slide show features generally show each photo at full-screen size, as in this example.

The slide show control toolbar in the upper left corner of Figure 12-2 is typical of what you'll see in the slide show features in photo editors. In addition to buttons to let you go forward or back by one slide, advance the slides automatically, and stop the automatic advance, there's also an Options button that opens a dialog box with a few basic options. You can, for example, specify the number of seconds between pictures when you have the automatic advance on.

If the photo editor you have doesn't have a slide show feature, that's no reason to do without slide shows. There are any number of programs you can get that will give you the same capability and more. One obvious choice is Microsoft PowerPoint. Among other benefits, PowerPoint can give you far more control over details than the slide show features we've seen in photo editing programs for things like adding a soundtrack or animating transitions—the change from one slide to the next—with special effects.

If you've ever used PowerPoint, or seen a PowerPoint presentation, you may think of PowerPoint slides as centering around text, but there's no reason why you can't ignore text and insert a photo—and nothing but a photo—in each slide. This is not the place to give a tutorial for using PowerPoint, but it is worth pointing out that inserting a photo in a PowerPoint slide is easy. What you want to do is insert a picture from a file. To do it, you choose the Insert menu, then chose Picture, then choose From File. You then find the picture on your disk, and double-click on it or highlight it and choose Insert. After that you can resize the photo, add text if you like, define the transition to the slide, add sound, and so on.

Of course, having a slide show on your computer is all well and good, but there's a good chance you don't keep your computer in the room where you plan to show the slides. The more useful choice may be showing the slides on your TV. We'll cover that next.

Viewing Your Photos on a TV Screen

The hardest problem you face for viewing your photos on a TV screen is deciding which of the myriad approaches you want to use.

- If your camera has a video output, you can connect your camera to your VCR or other TV equipment to show the pictures or record them for later viewing.

- You can use a PC-to-TV scan converter to connect the video output from your computer to your TV equipment for viewing or recording.

- You can use the more sophisticated approach of adding a video editing card (like the Matrox RT2500) to your computer. You can then use video editing software to create the slide show and record it on tape or optical disc with the recorder plugged into the video editing card.

- Finally, you can record your slide show to an optical disc, meaning one of the variations of CD or DVD writable discs. Once recorded, you can play it back on a DVD player attached to your TV.

We'll skip further discussion of the video editing card, on the grounds that you wouldn't get one just for creating slide shows, and if you have it for digital video, you probably already know how to use it. Here's what you need to know about the other three options. (Not so incidentally, don't skip the first section that follows, even if you're not interested in connecting your camera to your TV equipment. It includes some information that you'll need to know for the other options as well.)

Connecting Your Camera to Your TV or VCR

Your camera may or may not have a video output; check the manual if you're not sure. If it has one, it probably came with a cable that ends with connectors similar to the ones shown in Figure 12-3. One end of the cable should have either one or two RCA phono plugs, as shown on the left side of the figure. The other end, shown on the right side of the figure, will end in a plug that goes into the camera, and may vary from one camera to the next.

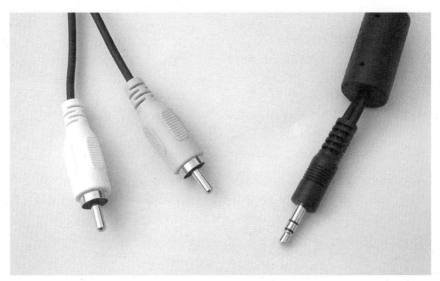

Figure 12-3 If your camera has a video output, it probably came with a cable similar to this one.

There's a possibility that the cable uses a different kind of connector instead of RCA phono plugs, notably an S-video connector, but whatever it uses should be a standard video connector. In any case, you'll need to make sure you have a matching connector on the video equipment you plan to plug into.

For this discussion, we'll assume you're using RCA phono plugs for the cable coming from the camera. If not, you'll need to modify our instructions to match the connector. We'll also assume that you have a VCR already hooked up using coaxial cable, with the input coming from an antenna, cable system, or the like, and the output going to your TV. This is probably the most common setup for TV equipment. However, the instructions will work with other combinations of video equipment also, with minor modifications. If you don't have a VCR, for example, you'll have to plug the cables into something other than your VCR, and you'll have to ignore our instructions for recording.

Before you do anything else, check your camera manual to see if you have a choice of video signals. Make sure the camera is set to NTSC, which is the acronym for the National Television Standards Committee, and is the standard for TV in the United States. (This, of course, assumes you're in the United States and using standard NTSC video equipment. The two other standards you may run across are PAL and SECAM. PAL is an acronym for Phase Alternating Line and is the dominant standard in Europe. SECAM is an acronym for Sequential Couleur avec Mémoire, and is the standard in France, most of Eastern Europe, and the Middle East.)

Lingo   *NTSC*, an acronym for the National Television Standards Committee, is the standard for TV in the United States.

Also check the camera manual to see if there is a special playback mode for TV. If there is, note how to set it. Most often, you'll use the same playback mode that you use to view photos on the built-in LCD. In addition, you may find a slide show mode that will play back all the images in the camera without your having to give any further commands.

■   To connect the digital camera to your TV equipment, first plug the RCA phono plug connectors into the video and audio *input* jacks on your VCR (you may have output jacks also, so be sure to read the labels).

●   Note that the plug and jack for the video signal are usually color-coded yellow. The plug and jack for the audio signal, which obviously will be available only if your camera can record sound, is usually color-coded white. In addition to being color-coded, the jacks on the video equipment are usually labeled with some variation of *Video In* and *Audio In*, as shown in Figure 12-4. If you have an AC power adapter for the camera, we recommend plugging that in too, so you don't use up your batteries watching the slide show.

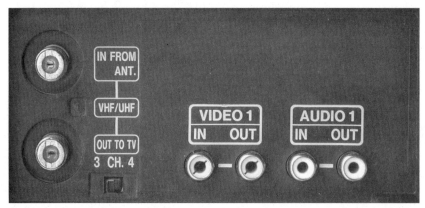

Figure 12-4 Look for RCA phono plug jacks labeled Video In or Video Input and Audio In or Audio Input.

■   Next, plug the other end of the cable into the camera, turn on the camera, VCR, and TV, and set your camera to the appropriate playback mode, optionally using the slide show feature, if there is one.

■ Now comes the tricky part. You need to set your VCR to use the video signal coming in through the RCA phono plug instead of the signal that's coming in from the coaxial cable.

● This is often well hidden in an onscreen menu system or elsewhere. In one VCR we tried this with, the only way to change the signal from one source to another was through a button on the remote labeled *WHO*INPUT*, which is something less than an intuitively obvious choice. With a little luck, you'll have the manual handy to get this information. At worst, you can contact the manufacturer and ask. (Alas, you may run into the same problem we did with one manufacturer, who responded that it was happy to hear that we had such an old model of TV still working, but it couldn't give us the information.) Once you know how to do it, set the VCR to use the signal coming from the camera.

■ At this point, you should see a photo on the TV screen, and be able to move from photo to photo, either by pushing a button on the camera or using the built-in slide show feature, if the camera has one. To record the photos on tape, simply put a tape in the VCR and push the VCR's Record button. Go through all the photos one by one, and you'll have a slide show on tape that you can play back any time you like.

Try This! If your camera has a video output, once you've successfully shown photos on the TV, you may find it can do a little more as a bonus. While still connected to the TV, switch the camera to picture-taking mode with the LCD active. In cameras we've tried this with, the camera feeds the live picture to the VCR, effectively turning the combination of camera and VCR into a camcorder.

Granted, it's a bulky, two-piece, highly limited camcorder. You can't move the combination very far, and—in the cameras we tried this with at least—the camera won't capture sound to pass on to the VCR. However, you can record pictures, and you could plug some other audio device into the VCR's audio to add sound. Despite the limitations, you might find this capability occasionally useful, and it certainly does no harm to find out if your camera can work this way.

Connecting Your Computer to Your TV or VCR

If you've created a slide show with your photos (or anything else for that matter) on your computer, and you want to record it on videotape, the simplest, least expensive approach is to use a scan converter. Basically, a scan converter plugs in between your computer's graphics card output and your VCR input to convert the computer graphics signal into the NTSC format that TVs and VCRs use. (At least, that's the format in the United States. We mentioned two alternatives in the section before this.)

Any scan converter that you get should come with complete installation instructions. Basically, however, all you have to do is plug in cables going from your PC graphics card to the scan converter and from the scan converter to the video input on your VCR. If the slide show includes sound, you'll also need to plug in a cable between the speaker output jack on your computer and the audio input jack on your VCR. In addition, you'll want to plug your monitor into the scan converter so you can easily monitor what's happening on the computer. Once you have everything plugged in, recording your slide show is as easy as pushing the Record button on your VCR and starting the slide show on your computer.

Prices for scan converters start at about $80 and run into four figures if you want a professional-level piece of equipment. For simply viewing photos, an inexpensive converter will do. There are a few things you need to check before buying however:

- Make sure the converter you get can convert from the resolution you normally use on your graphics card. Some inexpensive scan converters are limited to 640 × 480 resolution. If you normally use, say, 1024 × 768, make sure the scan converter can handle it.

- Also make sure it can handle the refresh rate you want to use, at the resolution you want to use. The vertical refresh rate, also known as scan rate, is given in hertz (Hz), which tells you how many times your screen is redrawn each second. Some scan converters are limited to 60 Hz (meaning that the screen is redrawn 60 times per second), which is slow enough to show a visible, headache-producing flicker on a computer screen. Unfortunately, if you don't already know the refresh rate you use, there is no easy way to check it that will work with all versions of Microsoft Windows, graphics cards, and monitors, but here are three ways to get the information, at least one of which will usually work:

  - First, check your monitor. Most monitors today have a built-in menu, usually called an on-screen display, or just OSD. If your monitor has one, look through it for a screen that will show you information about the current video signal. This should include two numbers, one given in kHz (kilohertz) and one in Hz (hertz). Ignore the kHz number. The vertical refresh rate is always in Hz.

● If that doesn't work, and you're using Microsoft Windows XP Professional, right-click anywhere on the desktop and choose Properties. Then choose the Settings tab, the Advanced button, the Adapter tab, and, finally, the List All Modes button to open the List All Modes dialog box shown in Figure 12-5. This should show the mode you are currently using highlighted as in the figure, including the vertical refresh rate in Hz. Once you know the rate you're using, choose Cancel repeatedly to back out of the dialog box and return to the desktop without making any changes.

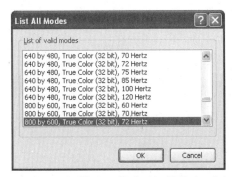

Figure 12-5 In Windows XP, the List All Modes dialog box shows you the current setting for your video card.

● If you're using Microsoft Windows Me or Windows 98, right-click anywhere on the desktop and choose Properties. Then choose the Settings tab, the Advanced button, and then the Adapter tab. You should see a Refresh Rate drop-down list, as shown in Figure 12-6. With a little luck, the refresh rate will be showing, as in the figure. If it's showing as Optimal or Adapter Default, you could open the drop-down list and choose a specific refresh rate, but unless you're sure you know what you're doing, we recommend against it. If you pick a rate that is too high, you could damage your monitor. In any case, once you've either gotten a rate or reached a dead end, choose Cancel repeatedly to back out of the dialog box and return to the desktop without making any changes.

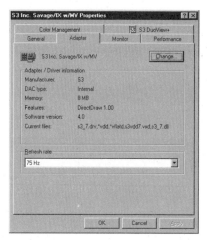

Figure 12-6 In Windows 98 and Windows Me, you may be able to find the refresh rate showing in a text box, like this.

- In addition to matching your video card for resolution and refresh rate, make sure the scan converter offers 24-bit color output. Watch out for claims that a scan converter is *compatible* with 24-bit color. That's all well and good, but you want to make sure it *delivers* 24 bits after the conversion. Some scan converters accept 24 bits but deliver less. That's acceptable for some purposes, but not for showing photographs.

- Make sure the converter has a place to plug in your monitor, so you can see the video on the monitor while you're working.

- Finally, when you buy the scan converter, make sure it either comes with the appropriate cables or that you buy the cables at the same time. Otherwise you'll have to take a trip back to the store or pay a second shipping charge.

Recording Your Slide Show on an Optical Disc

Recording a slide show on some variation of a CD or DVD is by far the most attractive option for moving your photos to your TV, at least at first glance: burn a disc, move it to a DVD player, and view the slides. Unfortunately, it's not quite that easy.

One drawback, of course, is that there are still a lot more VCRs than DVD players in the world at large—as witnessed by the relative numbers of tapes compared to discs at your local video rental store. But that's the least of it. The real problem is that there are all sorts of variations in both CD and DVD discs and drives that create compatibility issues.

One example should make the point. Any given DVD drive may or may not be able to read any given CD-R disc. Part of the problem is that the dye used in CD-R discs varies from one manufacturer to the next, and some of those dyes don't do a good job of reflecting the wavelength of laser light that DVD technology uses. In those cases, the information on the disc is essentially invisible to the DVD laser. Newer DVD drives and DVD players solve that problem by using two different lasers, with different wavelengths, for reading discs, switching the wavelength they're using as needed. Eventually all the older DVD players in the world will be replaced with newer models that can read CD-R discs, but for now there are still some DVD players around that simply won't read them.

Having pointed out that there are issues to worry about, we can still say that it's worth considering moving your photos to an optical disc, particularly if you plan to show them only on your own DVD player, and know that it can read the discs you produce. The various types of writable DVD drives (there are several) are still a rarity, so we'll assume that you're working with a CD-R or CD-RW drive for creating the discs.

The rule for moving photos to your TV by way of a CD-R or CD-RW disc and DVD player is simple: if your DVD player can read CD-R and CD-RW discs, and it can read Video CD (VCD) format—which is what you'll be using when you create the discs—it should be able to show your photos. Otherwise, it's anyone's guess whether it will or not.

If you haven't bought a DVD player yet, keep that in mind when you shop for one. If you already have a player and don't know if it meets these requirements, you can try creating a disc and see if the player can read it. Be aware, however, that even if the player can read CD-R, CD-RW, and VCD formats, it may still have a problem in particular cases, because of things like the choice of dye in a particular brand of CD-R disc. So if you have trouble getting it to read your discs, try at least two or three different brands of discs before you give up entirely.

Putting Slide Shows on CD

You can find a program for creating slide shows and writing them to a CD easily enough. Go to your favorite search engine (*www.altavista.com*, *www.google.com*, or *www.yahoo.com*, for example) and search for "cd photo slide show software" (without the quotation marks), then browse through the list it turns up, looking for sites that offer programs. We turned up several in a quick search, including Ulead DVD PictureShow (*www.ulead.com*) and Enterprise Corporation International's tvCD (*www.tvcd.biz*), which is shown in Figure 12-7. Both of these programs, as well as others, let you download a free trial version from the Web site, so you can try them before buying.

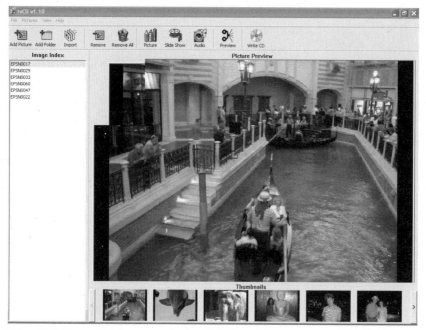

Figure 12-7 You can find programs on the Web like tvCD, which can create CDs that you can read in DVD players.

Using a program like tvCD is much like using a slide show program for viewing slides on your computer. In the case of tvCD, you can gather photos for the slide show by inserting them one at a time from anywhere on the disk, or by gathering all the slides you want to include in a single folder, and then telling the program to add everything in the folder to the slide show. You can easily put them in the order you want by renaming the files before importing them into tvCD so each one starts with a number. You then give a command to sort by file name. You can also rotate pictures as needed, add audio to the background, and specify a few basic settings like the resolution to use.

Where programs like tvCD differ from other slide show programs is by adding an option to write to a CD; you can see the Write CD command on the right-most toolbar button in the figure. Once you've created your slide show, and optionally previewed it on your system, you can put a CD-R or CD-RW disc in your CD-R/RW drive, give the Write CD command, and sit back while your computer creates the CD. (We tested the program with CD-R discs.)

When the program is done, you take the disc out of the drive, label it, and put it in your DVD player to view the slide show. That's all there is to it.

Before we leave the subject of optical discs, we should make one last comment about writable DVD formats. Even more than with different flavors of CDs, writable DVD compatibility with DVD players is problematic. It's a safe bet that compatibility will improve with time, but for the moment, the best we can say is that the industry is working on it. One designation you should be aware of is *DVD Multi*. A DVD drive with the DVD Multi logo is compatible with DVD-R, DVD-RW, and DVD-RAM discs, but it doesn't necessarily read DVD+RW or DVD+R discs. Keep that in mind if you're thinking about getting a DVD player or a writable variation of DVD for your computer.

Putting Your Photos on Your PDA

Hardly anybody realizes it yet, but the days of the wallet-size photo are numbered. The culprit is the color personal digital assistant (PDA). If you have a PDA with a color screen, a program that lets you view photos, and enough memory, you can carry a lot more pictures around with you on your PDA than you could ever carry in your wallet. Even better, they won't bend, fade, get dog-eared, or otherwise age.

Details for putting photos on your PDA vary, depending on the operating system and the particular PDA. We'll base this discussion on the Compaq iPAQ 3600 series Pocket PC, running Microsoft Windows CE 3, because that happens to be what we have. However, the general outline of what you have to do remains the same no matter what PDA you have.

- Start by finding out the PDA's resolution. This is a basic specification that should be in the manual or available on the manufacturer's Web site. In the case of the iPAQ, the resolution is 240 × 320 pixels.

- Crop and resample copies of the photos you want on the PDA to be equal to or less than the screen resolution. (We covered both cropping and resampling in Chapter 8, "Fun with Pictures: Basic Editing.") We strongly recommend that you leave the originals alone and work only with copies of the files, since you'll probably want to retain the higher resolution images for other purposes.

- Move the edited images into the folder on your hard disk that your PDA synchronizes with. For the iPAQ, the default name for the main folder is Pocket_PC My Documents, and the specific subfolder for pictures is called Pix. However, you might have given the main folder a different name when you installed your software.

- Make sure your synchronization program is set to synchronize files. With the iPAQ, the program in question is Microsoft ActiveSync. Details will vary depending on the program, but with the version of Active-Sync we use, you can check if it's set to synchronize files by choosing Tools from the ActiveSync menu, then Options. You then look for an entry called Files and make sure the Files check box is selected.

■ Synchronize your PDA. In the case of the iPAQ, you simply put the unit
   in its dock, and it will synchronize automatically.

Once the pictures are on your PDA, they're ready to view. With the iPAQ,
you open the main menu, choose Programs, and find the Picture Viewer icon.
Open Picture Viewer, and it will show you a list of files in the Pix folder. Tap on
the file name for the file you want to open, and your photo will show on the
screen. And the next time someone pulls a picture out of his or her wallet,
you're all set to counter by showing off your photos on your PDA.

Moving Photos from a Memory Card to Your PDA Memory

You can also move photos to your PDA by way of a memory card. This is worth
knowing about if the photos are on a different computer than the one you use
to synchronize your PDA with. The technique also works if you want to offload
some pictures from the memory card to free up space for more pictures.

Preliminaries first: We'll assume the pictures are already on the card, either
because you just took them with the camera, or you moved them to the card
using a card reader attached to your computer, or some equivalent approach.
We'll also assume you have an appropriate slot on the PDA that can accept the
card. Here again, details will differ depending on your PDA, but the basic
approach will be the same in any case. We'll base our comments, once again, on
the Compaq iPAQ 3600 series.

First, put the memory card in the PDA, and find a program that will let you
move files from the memory card to the PDA's main memory. With the iPAQ
3600, the program would be File Explorer. To open it, you first open the main
menu, choose Programs from the menu, find the File Explorer icon, and tap it.

Next, you have to navigate to the folder that has the files on the memory
card. Here's how you do that with the version of File Explorer we have handy:

■ File Explorer shows the current folder on the upper left of the screen,
   with a little triangle next to it indicating that it's a drop-down list. If you
   open this list, you'll see the entire path to the folder you're currently in,
   with My Device at the top level.

■ Tap on My Device to see all the first-level folder choices. If the storage
   card is in the PDA slot, one of the choices will be Storage Card.

■ Tap on Storage Card, and then navigate to the folder with the photos
   you want to move to your PDA. Tap on the photo you want to copy to
   the PDA memory to select it. Alternatively, if you want to copy all the
   photos, choose the Edit command at the bottom of the screen, and
   then Select All to select all the photos.

■ Next, tap and hold on one of the selected files to bring up the context-sensitive menu, and choose Cut. Navigate to the folder in the PDA's main memory that holds pictures for viewing. For the iPAQ 3600, you need to go to the My Documents folder, and then the Pix folder. Once there, choose the Edit command, and then Paste. File Explorer will copy the file to the Pix folder and delete it from the memory card. You can now remove the card and return it to your camera or elsewhere for safe keeping.

A Final Word on Viewing Photos on Screen

As we said at the beginning of this chapter, our intention was not to cover every variation possible for viewing photos on screen. However most other possibilities are variations on what we have covered. Moving a slide show to a writable DVD disc, for example, isn't very different from moving one to a CD-R or CD-RW disc. Similarly, the examples we gave of slide show programs and programs for writing to a CD are just that: examples of a much larger universe of programs. Some of those programs have more sophisticated features than others; some are easier to use. What we have tried to do here is give you enough information to get you started. We encourage you to explore further on your own.

Key Points

■ Many consumer-level photo editing programs, including the programs that come with many cameras, include slide show features, for viewing photos as slides.

■ You can also get slide show programs elsewhere, and even use sophisticated presentation programs like PowerPoint, to create slide shows.

■ If your camera includes a video output, you can connect it to your TV to view the photos, or to your VCR to view or record the photos.

■ The hardest part of connecting your camera or computer to your TV equipment may be finding the control for switching to the port the camera or computer is connected to. This is often hidden in onscreen menus, or available only through an obscure button on the remote control.

■ To connect your computer to TV equipment, you need a scan converter. An inexpensive converter will do, but make sure it supports the resolution and refresh rate you use, and that it has a place to plug in your monitor.

- Recording a slide show on optical disc can be problematic because of incompatibilities between different formats, but it's potentially the easiest way to show your photos on your TV.

- If you plan to record your slide show on CD-R or CD-RW, make sure the DVD player can read CD-R and CD-RW discs as well as VCD format.

- You can find an assortment of programs for putting slide shows on CD-R and CD-RW discs by searching the Web.

- You can also put your photos on your PDA, if it has a program for viewing photos. First edit copies of the photos to match the PDA screen's resolution, then move them to the folder that the PDA uses for synchronizing files.

- If the PDA can accept your camera's memory card and has a program for moving files from one folder to another within the PDA, you can also move photos to the PDA memory by way of a memory card. Put the photos on the card, put the card in the PDA, and then cut and paste the files from the card to the PDA's main memory.

Chapter 13

Sharing Your Photos: E-mail, Letters, and Web Sites

If you're like most people, you probably want to share your photos—with friends and family, clients and customers, and maybe the world at large. Digital cameras in concert with e-mail, computers, and the Internet let you share in ways that simply weren't possible just a decade or two ago.

Want to send a picture of your new baby to various friends who live all across the country—or in different countries, or different continents for that matter? You can send it to all of them all at once as an e-mail attachment.

Putting together a monthly company newsletter for your clients or personal holiday newsletter for your friends? Add photos to help illustrate what you're talking about.

Want to put your photos on the Web where anyone can see them from literally anywhere in the world? Easy. You don't even have to have your own Web site or learn how to create one. There are services that will let you create online albums for free; they'll even give you the choice of making the photos available for anyone at all to see, or just the people you give a password.

You probably already knew about most, if not all, of these possibilities. If you haven't taken advantage of any of them, it may be because you think they are hard to do. They're not. We promise.

E-mailing Photos

If you have a digital camera, you probably have a computer too, and you very likely have a connection to the Internet. If so, you already have e-mail as well, whether you use it or not. (If you don't have a computer and Internet connection, this section may inspire you to get them.)

If you don't have e-mail yet, or have it but don't use it much, you may not know how to add a photo to a message. If you use e-mail enough to know how to attach a file to a message, you already know how to send a photo, but you may not know the best way to send it, which is mostly about keeping your photos to a reasonable size. Let's tackle that part first.

What's a Reasonable File Size?

Reasonable size is a flexible term. Well into the early 1980s, the modem speed that most people used was 300 bits per second (bps) for both sending and receiving. Sending or receiving a 500 KB file at that speed would take an absolute minimum of 4.6 *hours*—and actually more in the real world. We rounded numbers down for a quick, back of the envelope calculation, and this assumes everything is working perfectly, without any bursts of noise that require any retransmissions of data. Most people would not consider that a reasonable amount of time for a file transfer, which would make a 500 KB file an unreasonable size in 1980.

Today's modems are much faster, and you can usually count on a *minimum* reliable speed of 28,800 bps—almost 100 times as fast as early modems. Sending a 500 KB file at that speed takes about 3 minutes, making 500 KB a reasonable size for sending, and reasonable for receiving, as long as the person who is getting the file actually wants it. If you're sending a photo of your one-day-old baby to his grandmother, three minutes isn't all that long to wait on either side. If you're sending the same picture to all your friends as a baby announcement, however, some people on the receiving side will almost certainly perceive the three-minute wait as far too long to wait for one e-mail message.

If you're using a broadband connection like cable or Digital Subscriber Line (DSL), the definition for reasonable size changes again. The minimum speed you likely have for sending files is 128,000 bps, which cuts the time for sending a 500 KB file to less than a minute. Even better, your speed for receiving data is likely in the neighborhood of 500,000 bps or better, which cuts the time for receiving the file to about 10 seconds. At that speed, few people, if any, would consider the file size unreasonable.

Our point is that what counts as a reasonable file size depends in large part on both the connection you have for sending the file and the connection on the other side for receiving it.

Tip A file will take roughly 35 seconds per 100 KB to send or receive by modem at 28,800 bps. Many people will start getting impatient if they are waiting for other messages and the time starts approaching a full minute for a message they are not expecting.

The first thing you should do before sending photos larger than about 150 KB to friends and clients by e-mail is find out what kind of connection they have. If you make a habit of sending multimegabyte files to people who have only a modem connection, they may not be friends or clients for long. Even for people with broadband connections, it's a good idea to ask first if they want to see the photo. If it's not possible to ask—because you're sending the photo to a long list and don't want to check with everyone—assume that at least some people on the list have modems, and choose a file size accordingly.

Watch Out for Low Ceilings

Even if you are sending a file to someone with a broadband connection who very much wants a high-resolution version of a photo and doesn't mind getting an 8 MB or 10 MB file, you may not be able to send it.

Some e-mail systems limit the size of files you can send or receive. One of us, for example, has one e-mail account that automatically strips off any file larger than 1 MB attached to an e-mail message. Send a larger file to that e-mail address, and the file simply disappears into the ether without any warning to either the sender or receiver. Try sending it from that address, and the system rejects it. Check with whoever provides your e-mail—usually your Internet service provider (ISP)—to find out if there are any limits on file size. If you need to send a large file, and it's important that the photo get through, it's a good idea to ask the person who's expecting it to check on his or her end also.

There's another issue that can affect the maximum file size you can use too. As we've mentioned elsewhere in this book, most commercial e-mail systems provide a specific, limited capacity for your inbox, the combination of all inboxes on your account plus any Web site you set up, or both. If you send a 5.1 MB file to someone whose ISP allows a firm maximum of 5 MB, the system will reject the message. If you send a 4.9 MB file, it may get through, but only if the inbox is essentially empty. If you send three 2 MB files, each as part of a separate e-mail message, the first two may get through, depending on how full the inbox is, but the third will get rejected.

The moral here is to find out the limits you have to deal with and choose your maximum file sizes accordingly.

Keeping Files to a Reasonable Size for E-mail

The good news is that it's easy to keep file size down, and there's often no reason to send a large file in any case. There are also some ways to work around certain limits. Here's a short checklist of what you can do to minimize file size and get the file through:

- Don't use a larger resolution than you need.

    - For viewing on screen: as we've discussed elsewhere in this book, files are best viewed on screen using one pixel in the file to one pixel on the screen. In most cases for viewing on screen, 800 × 600 resolution is all you need. In many cases, 640 × 480 is enough.

    - For printing: as we've also discussed elsewhere in this book, you'll get high-quality photos at 200 pixels per inch (ppi), and for most purposes, 150 ppi is all you need. Beyond that, higher resolutions add little or nothing to image quality. If you're sending a photo to be printed, find out what size the print will be, and resample the photo accordingly. (We discussed resampling in Chapter 8, "Fun with Pictures: Basic Editing.")

- Compress the file as much as you can. Use JPEG compressed format unless you have a compelling reason not to, and consider using the highest compression available in your editing program to make the file even smaller. (We discussed compression in Chapter 3, "Getting Started with Digital Photography," in the section "Choosing Resolution and Compression Settings.")

- If you're sending more than one photo, attach each one to a separate e-mail message. The more files you attach to the same message, the more chance that you'll bump against a system limit. Also, if a poor connection prevents a message from getting through the first time, it has to be retransmitted. By attaching each file to a separate message, you ensure that if a burst of noise prevents one file from getting through, you won't have to resend all of them.

■ If you're sending enough large files so there's a chance that you'll fill the receiver's inbox, coordinate to make sure he or she receives each message and empties the inbox before you send the next one.

■ If you must send a high-resolution file that isn't compressed, or isn't compressed very much, consider using a different approach, like putting the files on a CD-R disc and physically mailing the disc.

The Mechanics of E-mailing a Photo

If you already know how to attach a file to an e-mail message, feel free to skip this section; you probably won't find anything new in it. However, if you don't know how, and are a little hesitant to try, you should find this section helpful.

Details for sending photos vary slightly with specific e-mail systems and programs, but the concepts remain the same from one system to another. For this discussion, when we use a specific program as an example, we'll also give you suggestions for finding equivalent features in other programs.

First, we need to talk a little about e-mail.

E-mail Basics

E-mail doesn't work all that differently from regular mail, particularly as regular mail works in rural locations. If you live in a rural location, you have a mailbox with a flag you can raise. If you have incoming mail, the mail carrier will put it in the mailbox for you to pick up. To send mail, you can leave the mail in the mailbox and raise the flag, so the mail carrier will know to stop and pick up the mail.

E-mail works just like that. For simplicity, first think of a simple e-mail system that lets you send messages just within a small company. The system has two pieces. The *server* is the equivalent of your local post office. It runs on some system somewhere in the company; you don't need to know where. Its job is to receive e-mail and make it available when someone asks for it. The *client*, which usually runs on your individual computer, is the equivalent of a rural mailbox. You receive messages in the inbox part of the client and put messages in the outbox part of the client to send them. When you give a send or receive command from the client, it's the equivalent of telling the mail carrier that it's time to do his or her rounds.

Lingo   An *e-mail server* is the equivalent of a post office in the real world. An *e-mail client* is the equivalent of a mailbox.

Now consider what happens with the Internet. If you attach directly to the Internet, whether by modem or broadband connection, your ISP's e-mail system is your e-mail server. Just like your local (actual) post office, it's part of a network that lets it send mail to and receive mail from any other post office. In this case, however, the other post offices are other e-mail servers connected to the Internet. The only difference between the two situations is that the post office network physically moves the mail in trucks, planes, and trains. The e-mail network moves bits.

Some companies have their own e-mail servers. If your e-mail goes through a company server, the situation is essentially the same, except that your company's server is the local post office, your client deals with your company server, and the company server deals with moving mail to and from the Internet.

And that's basically how e-mail works. For a little more on the subject, take a look at the sidebar that follows.

E-mail: Variations on a Theme   If there's any variation on e-mail that you can think of, somebody is probably providing it.

In most cases, the inbox and outbox are physically on the same computer that you're using to view them. Incoming e-mail messages actually get moved to that computer, and outgoing messages are created on the computer, stored there, and sent from there. In some cases, however, the inbox and outbox—and the messages in them—stay on the server. The client has to retrieve the message from the server each time you want to read the message, and the outgoing messages are saved on the server when you create them.

This is most obvious with Web-based e-mail. The client isn't installed on your computer; it's a Web-based application that you run by way of your browser. What's more, your incoming messages aren't sent to your computer, and your outgoing messages aren't saved there. Instead, you run the client by going to the Web site, where you can read your mail, or create new outgoing mail, with all the messages stored on the Web site.

Just to confuse matters, some Web-based e-mail sites offer the option of letting you send and receive e-mail using a standard client, with the messages stored on your local computer. In perfect symmetry, some ISPs that you would normally expect to use with an e-mail client running on your computer also let you read and create e-mail through a Web-based client. In all of these variations, however, the basics remain the same: the server provides a post office, and the client provides an inbox and outbox.

Sending a Photo Using an E-mail Client

There are any number of e-mail clients available that work with almost any e-mail system. The vast majority of e-mail servers use one of just a few standard protocols for moving messages to and from clients and other servers, and most clients support all the common protocols. The glaring exception to this rule is America Online (AOL). AOL uses a proprietary protocol, which means you can't use it with standard e-mail clients. Even with AOL, however, the basics for sending photos are the same.

For this discussion, we'll use Microsoft Outlook Express 6—the version that comes with Microsoft Windows XP—in our examples. If you use a different client, details will differ, but the basic steps won't. We'll also assume that you already have your e-mail client installed and working. If you don't, contact your ISP's technical support number to find out how to set up your client for your ISP.

> **Note**   Many e-mail clients will let you insert photos and other graphics in the message itself, but other e-mail clients will strip out those inserted photos when they receive the message. If you want to be sure the photo will get through, send it as an attachment.

■   Start by making sure you know where the file that you want to send is, and how to find it. If you've stored it in a photo album program, you may need to export it from the program folder into a Windows folder.

■   Load your e-mail client, or go to the appropriate Web site if you use Web-based e-mail, and start an e-mail message. In Outlook Express, shown in Figure 13-1, you can simply click on the Create Mail button at the extreme left of the toolbar. A typical option for Web-based mail would be Compose.

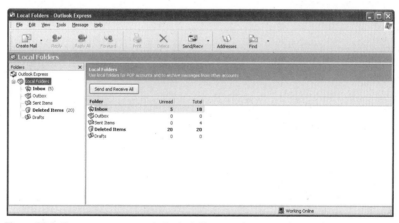

Figure 13-1 To create a message in Outlook Express, choose the Create Mail button on the toolbar.

■   The window for creating the e-mail message should look something like the one in Figure 13-2. Your e-mail address will show in a From text box, and you should be able to enter as many e-mail addresses as you like in a To or CC text box, or both. Typically you separate the addresses with commas. (AOL uses a separate line for each address.) Write any message you like in the message area.

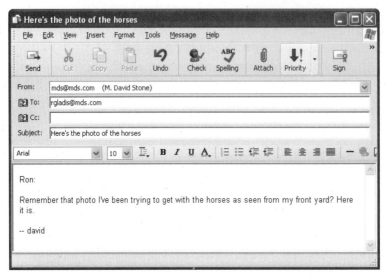

Figure 13-2 The window for creating a message should look much like this one in any e-mail program.

- To attach the photo, look for a command with a name like Add Attachment or Insert Attachment. In Outlook Express, you can choose the Insert menu, then File Attachment, or you can click on the Attach button on the toolbar. In most e-mail programs, you'll find an equivalent button, usually with a paper clip icon on it, as in Figure 13-2.

- At this point in Outlook Express—and almost any other program including Web-based e-mail—the client will open a dialog box that you can use to move to the folder with the photo you want to send. Find the file, highlight it, and choose Attach or the equivalent command in the program you're using.

- After you attach the file, you should see some indication that there's a file attached. Figure 13-3, for example shows a newly added text box, labeled Attach, with the file name in the text box.

Figure 13-3 Note the newly added text box labeled Attach, complete with a file name.

■ All that remains is to close the message and send it. In Outlook Express, you can choose the Send button on the toolbar above the message text. Depending on the program you're using and how it's set up, this may send the message immediately, or it may put it in the outbox for sending later.

■ If Outlook Express is set up for the message to go to the outbox, you can then actually send it from the outbox by clicking on the Send/Recv button in the main Outlook Express window, or by choosing Tools, then Send and Receive, and choosing the appropriate option from the submenu that opens. If you connect by modem and aren't currently connected, Outlook Express will then dial your ISP, make a connection, send the message, and retrieve any messages waiting for you. If you connect by broadband and are permanently connected, it will simply find your e-mail server on the Internet, and then send and receive the messages.

Defining an E-mail Account   Your ISP should give you the information you need to get
started with e-mail. However, some ISPs limit their help to specific e-mail clients. If you want to use
a different client, here's what you need to know.

Briefly, you need several pieces of information that you can get only from your ISP:

- Your user name

- Your password

- Your e-mail address

- The type of incoming mail server (meaning what standard it uses; the most common is
  Post Office Protocol 3, more often referred to simply as POP3)

- The name of the incoming mail server to use

- The name of the outgoing mail server to use

Armed with this information, you can define the account in the client. Look for an option to
add an account, then go through the settings, plugging in each one that you have. Some programs,
including Outlook Express, use a wizard to walk you through the steps of defining an account, but
they may skip some items. Be sure to look at the settings when the wizard is done, and make sure
that all the information you got from your ISP is, in fact, filled in.

One other thing: you also need to know whether you need to change anything from its usual
defaults, and, if so, what. This may be a little harder to get for programs that the ISP doesn't support.
Try using the default settings first. If they don't work, call your ISP's technical support number and
ask about each setting you can find in the program, whether you know what the setting does or not.
If you don't get answers the first time, try calling again later. You may get someone else who knows
more, or is simply more willing to be helpful.

Adding Photos to Documents

Adding a photo to a document is usually as easy as finding a command in the
program to insert a picture or photo. For example, in Microsoft Word 2000,
which we happen to be using to write this book, if you want to insert a picture
from a file, you open the Insert menu, choose Picture to open a submenu, and
then choose From File. You then navigate to the photo file, highlight it, and
choose Insert.

Almost any program that lets you insert photos will have a similar command. If it doesn't, you can resort to copying and pasting the photo. Open the
photo file in any program that will let you see it, select the photo, and give the

copy command from within that program (choosing Edit and then Copy will usually work). Then open or move to a window with the document you want the photo in, position your cursor where you want the photo, and give the command to paste (choosing Edit and then Paste will usually work).

There are variations on this procedure that you should be aware of. For example, any given program may let you insert a link to a photo instead of inserting the actual photo. If you then edit the original photo, you don't have to insert the new version into the document. You can simply open the document and have the program read the changes from the original photo file that the document is linked to. Alas, details for linking vary greatly from one program to another. If you're interested in linking to a photo file, you'll have to investigate the options in the program you're using.

Note   The comments we made about file size in the section "Keeping Files to a Reasonable Size for E-mail" apply just as much to photos you plan to insert in a document—particularly if you plan to e-mail the document to others.

However you insert a photo, there are some things you should know about how to work with a photo in a document. For this discussion, we'll assume you inserted the photo in a word processing document, which is the most likely destination. We'll use Word 2000 for our specific examples, but the concepts apply to working with photos in most word processing programs—and many other kinds of programs as well.

Programs can treat photos in one of two basic ways. They can insert them *in-line*, which translates to inserting the photo in a line of text, and treating it as a single large text character, or they can insert them in a way that lets them *float*, so you can move them anywhere you like on the page.

Lingo   Programs can insert a photo *in-line*, and treat it like a large text character, or in a way that lets it *float*, so you can move it anywhere on the page.

Inserting a photo in-line works well when you want the photo to take up the entire width of the text column, but it's a problem when you want the photo to take up only part of the width, with the text flowing around it. As you can see in Figure 13-4, if you make the photo smaller than the width of the column, you get only one line of text to the side of the photo—in the same text line as the photo—and a lot of white space.

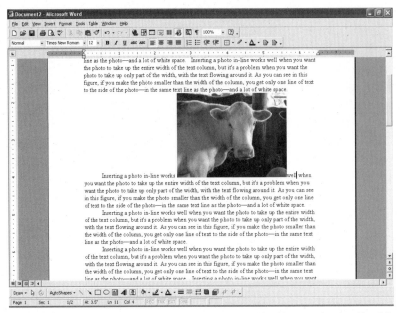

Figure 13-4 Insert a photo in-line, and you may wind up with white space to the side of the photo.

Inserting a photo so it can float anywhere on the page solves that problem. We've inserted three small versions of the same photo in Figure 13-5 to show that you can position the photo in different ways to have the text flow around it differently.

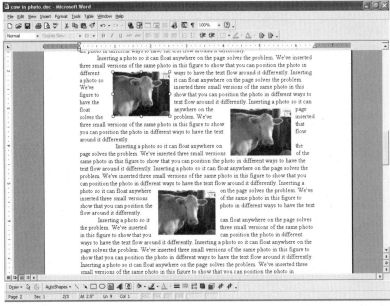

Figure 13-5 Most programs also let you insert a photo so it can move anywhere on the page, with text flowing around it.

In Word, you can redefine any individual photo to make it in-line or floating, and you can also modify some settings for a floating photo. To select the photo and open the Format Picture dialog box shown in Figure 13-6, double-click on the photo, then choose the Layout tab. As you can see in the figure, you can adjust whether the text flows around the photo, prints over the photo, or is hidden behind the photo. The Advanced button opens a dialog box that gives you more control over some of the details, like whether to wrap text on one side or both sides of the photo. Any given program may or may not have equivalent features.

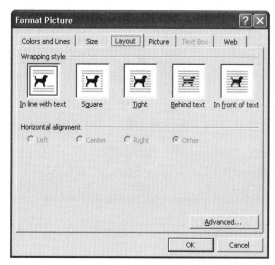

Figure 13-6 In some programs you can change between floating and in-line formats for any given photo.

One other option you should know about is defining the photo as in-line, but putting it inside a table cell. This is particularly helpful in layouts like the one in Figure 13-7, where you may want to put two photos in one column, one taller photo in a second column, and text describing the photos in a third column. We've left the grid lines for the table showing so you can see what we did, but you can turn some or all of them off for a better looking layout.

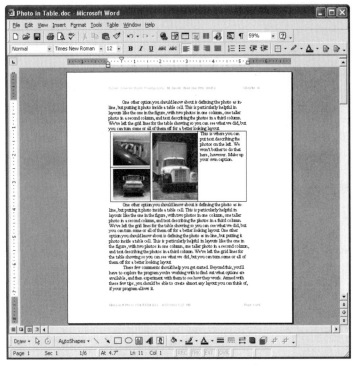

Figure 13-7 We created this layout by inserting a three-column, two-row table, merging the two cells in two of the columns, then inserting the photos as in-line photos.

These few comments should help you get started. Beyond this, you'll have to explore the program you're working with to find out what options are available, and then experiment with them to see how they work. In most programs you'll find that you can create almost any layout you can think of.

Sharing Photos on Photo Web Sites

In Chapter 5, "Special Issues for Digital Photography" in the section "The Third Way," we mentioned that there are some Web sites, like *www.photoaccess.com*, that let you store your photos online for free. They actually let you do much more than that.

PhotoAccess, for example, lets you store photos, order prints, and order all sorts of merchandise with your photos printed on them—from t-shirts to gift wrapping paper to jigsaw puzzles. Figure 13-8 shows the choice for a coffee mug, for example. The site will even record your photos to an archival CD.

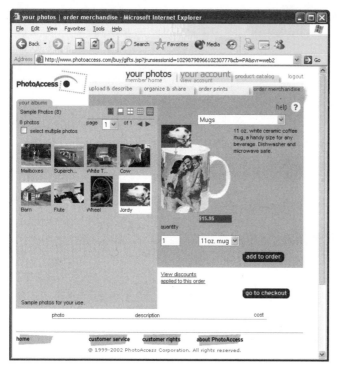

Figure 13-8 You may be amazed at what you can order online with your own photos emblazoned on it.

As we mentioned in Chapter 5, we have some reservations about depending on a Web site to store all your photos. Aside from privacy issues, we simply don't feel comfortable depending on a Web site to maintain our irreplaceable photos. If the Web site fails as a commercial enterprise, there's no guarantee that you'll ever see your photos again. Even if the Web site does exceedingly well, there's no guarantee that it's backing up your photos properly and won't lose them if a hard disk crashes. Our advice is simple: if you store photos on a commercial Web site, make sure they are copies of photos that you have stored elsewhere as well.

Having said that, we'll also point out that these Web sites are not only a good place to get your photos printed on oddball items like jigsaw puzzles and coffee mugs, they are also a great boon if you want to share your photos on the Web and don't want to maintain your own Web site.

Figure 13-9 shows the PhotoAccess page for organizing your photos into albums and sharing them with friends, with the Share options showing. We won't take you through the procedure step by step because it's explained well on the Web site. (For details on how to use the feature, choose the Help option at the top right of the Web page.) However, we want to point out one nice touch. The Send Me A Link button lets you ask for an e-mail with the link to your shared album so you can forward the information to anyone you like. If you've password protected the album, the e-mail will also include the password to forward along with the link.

Figure 13-9 Some Web sites, like this one, let you create online albums to let others see your photos on the Web.

How to Find More Sites for Sharing Photos

Here again, as with our discussion of inserting photos into documents, this is meant strictly to get you started. You'll want to explore other features on the PhotoAccess site, and also look at some competing sites. We found a fairly large number of sites by going tob your favorite search engine (*www.altavista.com*, *www.google.com*, or *www.yahoo.com*, for example) and searching for "store share photos" (without the quotation marks).

Other Web Sites of Interest

We'll end this book in an open-ended way, by pointing out that there are any number of Web sites you may want to explore. Some, like *www.photoaccess.com*, offer products or services that you may find useful. Others will help you learn more about digital photography and keep current on new products and other developments, with reviews and technical information. Here's a short list of just some of the Web sites that are worth taking a look at:

- *www.clubphoto.com*: order prints; share and store photos online
- *www.image-edit.com*: photo restoration; can restore "irreparable" photos
- *www.imaging-resource.com*: reviews, discussions, and educational materials for taking better pictures, and more
- *www.ipix.com/showcase/*: a collection of 3D immersive images created with iPIX technology
- *www.megapixel.net*: a monthly Web magazine on digital cameras
- *www.steves-digicams.com*: reviews, discussion forums, and news

You might also want to try searching for any one of these sites on your favorite search engine, to find other sites that talk about them. Those other sites will probably mention still other sites to look at. Once you get started, you shouldn't have any trouble finding more sites. There are plenty of them out there just waiting for you. Have fun.

Key Points

- How large a file is reasonable to send by e-mail depends on the speed of the connection and how much the recipient wants it.
- A 500 KB file will take a bit under three minutes to send or receive over a 28,800 bps modem, but only about 10 seconds or less to receive over a cable modem or DSL connection.
- Some e-mail systems put a limit on file size, and strip off anything larger than the limit.
- Most e-mail systems put a limit on the size of an inbox, or the total of all the inboxes and other disk space used for one account. Once the limit is reached, the system will reject additional messages, until you make more room by receiving the message to your computer, and deleting it from the e-mail system.
- To keep your files to a reasonable size for e-mail, use compression and don't use a larger resolution than you need.

- If you're sending multiple photos, attach each one to a different e-mail message.

- All e-mail systems are variations on a theme. A server application acts as a post office. A client application acts like an actual mailbox. All the servers on the Internet communicate with each other to move e-mail around. Your client communicates with the server you use, such as your ISP's e-mail server.

- To send a photo, open your e-mail client, enter the e-mail address for the person you're sending the photo to, attach the photo, and send the message.

- Most word processors and many other programs include a simple command to insert photos. In most cases you can also copy the photo from a photo editor and paste it into a document in another program.

- Photos can be formatted to be in-line, in which case they are treated like a text character, or formatted to float so you can move them anywhere on the page. If you're working in a program that lets you format the photo as floating, you should also be able to define whether text flows around the photo.

- Some online systems let you store your photos online as albums and let other people see them. You may have the choice of using a password to prevent people without the password from seeing your pictures.

Index

M. David Stone

M. David Stone is an award-winning writer, a contributing editor at *PC Magazine*, and a onetime semi-professional photographer. He has been working with and writing about personal computing since 1981, with imaging technologies—cameras, printers, monitors, and scanners—an area of special interest and recognized expertise. David has written more than 2,000 articles in national and international publications. In addition to *PC Magazine*, his work has appeared in *Wired*, *Computer Shopper*, *Windows Sources*, *PC Sources*, *InfoWorld*, *ExtremeTech*, and *Science Digest*, where he was Computers Editor. He has also written a column for the *Newark Star Ledger*, and has written more than ten books, including, most recently, *The Underground Guide to Color Printers*, published by Addison-Wesley, and *Troubleshooting Your PC*, published by Microsoft Press.

Ron Gladis

Ron Gladis is a reformed electrical engineer who switched to a career in the visual arts. Ron's initial primary interest was fine art photography, which he exhibited regionally and abroad over a span of 15 years. He's also taught and guest lectured on photography as a fine art at the University of Pennsylvania, Temple University, the Philadelphia College of Performing Arts, the Philadelphia Art Alliance, and the Academy of Fine Art, and he spent a year as Photographic Artist in Residence at Widener University in Pennsylvania. Currently, Ron is Principal and Director of Gladis and Gladis Pictures, an award-winning video and film production company, which he founded in 1988. The company, affectionately known as G2 in the industry, creates film, video, and computer-based multimedia training programs for major corporations.

The manuscript for this book was prepared and submitted to Microsoft Press in electronic form. Pages were composed by nSight Inc. using Adobe FrameMaker+SGML for Windows, with text in Garamond and display type in ITC Franklin Gothic Condensed. Composed pages were delivered to the printer as electronic pre-press files.

Cover Designer:	Tim Girvin Design
Interior Graphic Designer:	James D. Kramer
Compositor:	Donald Cowan
Project Manager:	Julie B. Nahil
Copy Editor:	Teresa Horton
Technical Editor:	Thomas Keegan
Proofreaders:	Jacqueline Fearer, Katie O'Connell
Indexer:	Jack Lewis, J&J Indexing

Get a **Free**
e-mail newsletter, updates,
special offers, links to related books,
and more when you
register on line!

Register your Microsoft Press® title on our Web site and you'll get
a FREE subscription to our e-mail newsletter, *Microsoft Press
Book Connections.* You'll find out about newly released and upcoming
books and learning tools, online events, software downloads, special
offers and coupons for Microsoft Press customers, and information
about major Microsoft® product releases. You can also read useful
additional information about all the titles we publish, such as detailed book descriptions, tables of contents and indexes, sample
chapters, links to related books and book series, author biographies,
and reviews by other customers.

Registration is easy. Just visit this Web page and fill in your information:

http://www.microsoft.com/mspress/register

Microsoft®

- -

Proof of Purchase

Use this page as proof of purchase if participating in a promotion or rebate offer on
this title. Proof of purchase must be used in conjunction with other proof(s) of
payment such as your dated sales receipt—see offer details.

Faster Smarter Digital Photography
0-7356-1872-0

CUSTOMER NAME

Microsoft Press, PO Box 97017, Redmond, WA 98073-9830